A FRESH LOOK AT FIGS

Traditions, Myths and Mouth-Watering Recipes

Pamela Allardice

HILL OF CONTENT

First published in Australia 1993
by Hill of Content Publishing Company Pty Ltd
86 Bourke Street, Melbourne Australia 3000

Illustrated by Simone Bennett
Typeset by AdverType, Albert Park, Victoria
Printed and bound in Singapore by
Kyodo Printing (Singapore) Pte Ltd

National Library of Australia
cataloguing-in-publication data

Allardice, Pamela, Date-
A fresh look at figs.

Bibliography.
ISBN 0 85572 225 8.

1. Fig. 2. Cookery (Figs). I. Title.
641.3437

Contents

Outside the courtyard, near the entrance gates, is a great garden of four acres with a fence running around on either side. Here grow tall flourishing fruit trees: pears, pomegranates and shining apples, sweet figs and luxuriant olives. The fruit of these trees never falls off or fails, but is there winter and summer, all the year round; for the west wind, always blowing, forms new fruit while it ripens others. Pear after pear comes to maturity and apple after apple, one bunch of grapes after another, and fig after fig. There too, beyond the furthest row of vines, are well kept beds of herbs which are fresh all the year round, and there are two springs, one of which sprinkles the whole garden, while the other is channelled under the entrance of the courtyard to issue by the lofty palace and here the townsfolk draw their water. These were the splendid gifts of the gods in the house of Alcinius.

Homer, *The Odyssey*, Book vii

Who gives a fig?

Luscious fresh figs are the essence of summer. They evoke lazy, sun-filled days with their delicious fruit glistening in the heat in green and smoky purple—lovely to look at and a joy to eat.

Figs are also perhaps the most sensual fruits, with their blooming, bursting skins and lush flesh. Fresh figs are appealing , sweet and mellow to eat when they are fully ripe and soft and their attractive shape and flavour make them perfect both for desserts and for savoury dishes. Dried or fresh, figs are popular, preserving the incomparable flavour of their summertime beauty year-round.

Our ancestors recognised the value of figs, too. Figs were popular everywhere in the Near East at least as early as 3000 BC. Assyrians used fig syrup as a sweetener, figs were grown in the Hanging Gardens of Babylon and were buried in hot desert sands to dry for preservation. Even before domestication, figs grew along the banks of the Jordan River and around the Dead Sea. Remains of figs, dating back to Neolithic times, were found in excavations at Gaza. In Palestine, along with dates, figs are thought to have formed the basis of most peoples' diets and were used either fresh

Now, sing of the fig, Simiane,
Because its loves are hidden.
I sing the fig, said she,
Whose beautiful loves are hidden,
Its flowering is folded away
Closed room where marriages are made;
No perfume tells the tale outside.
Andre Gide

or dried. When people travelled on long journeys or were besieged during battle they could—and did—subsist on dried figs.

The Leningrad Papyrus from Egypt, dated 1800 BC, includes a story of a sailor saved from a shipwreck in the Red Sea, off the Sinai Coast, by being thrown up on an island where he found wild figs, grapes and other fruits. Other ancient Egyptian records list dried figs, oil, nuts and honey as being items imported from Syria. The Egyptians were thought to have been quite preoccupied with their digestions and so had a habit of requiring regular fasting and purging schemes as part of their cleansing rituals. This is one reason why figs—with their mild laxative properties—appealed to them as a food which was not only delicious but also good for them.

The Egyptian Pharaohs ate figs when the Sphinx was young. Hieroglyphs praised the value of dried and fresh figs as food and murals depict their husbandry as long ago as 2400 BC. We know from Egyptian temple paintings that baskets of dried figs accompanied a Pharaoh on his journey to the afterlife and paintings show the very finest of figs being selected and saved for the King. We also know that figs may have been cultivated in an espaliered form in the elegant walled gardens which formed part of the Pharaoh's court, for we have descriptions of figs 'climbing' on walls which faced south.

The Greek playwright Menander is said to have described himself as a simple man with the phrase, "I call a fig a fig, a spade a spade." However, the definition of figs is not quite so simple a matter. There are figs which ripen underground and others which grow high in the air, strangling and even killing their hosts. There is even a 'sea fig', or 'sea squirt', a wineskin-shaped creature which attaches itself to rocks on the bottom of the Mediterranean sea! Then there is the Peepul or *Bo* tree (*F. religiosa*) which is often planted near Indian Buddhist temples. The Buddhists believe that, while sitting under this tree, their spiritual leader gained enlightenment. Then there is the Sycamore fig (*F. sycomorus*) —also known as the Egyptian fig or the Pharaoh's fig; the wood of this tree is valued for its hardness and resistance to

Catherine brought me up as a great treat yesterday at dinner, ham dressed with as much garlic as could be stewed into it, and a plate of raw figs, telling me I was to eat them together!

John Ruskin, 9 July 1874, Assisi Italy

rotting and, in ancient Egypt, was used to make sarcophagi. Another interesting fig is the South East Asian *F. elastica* which, as its name suggests, emits a type of thick white latex which can be used to produce a sort of rubber.

The fig is a member of the mulberry family. The main type of eating fig is the Common Fig, or *F. carica*. This is one of the oldest known plants and is thought to have been a native of Western Asia and the eastern Mediterranean region. It is still found in the Caucasus, where it is considered to have originated in the forests in the deep valleys and gorges of the Tajik Mountains. Its succulent fruit, both fresh and dried, has been valued since earliest days. The exact lineage of this ancient plant is quite difficult to trace.

Even today there is occasionally some confusion, simply because one variety of fig is often given several names which vary from country to country.

There are over 700 fig varieties, the Smyrna fig from western Turkey probably being the best known. The thin-skinned, soft kind known as "Eleni" are perhaps the most valued type of Smyrna fig. Another delicious one is the Calimyrna fig (*Cali*fornia and *Smyrna* cross) which is cultivated both for use as a fresh fruit and for drying. Maltese figs are also very good while the Greek ones, being thicker-skinned, tougher and containing less pulp, are regarded as slightly inferior.

Figs are usually classified by their colour—green, yellow, black, purple or red—or even by their shape—large, small, round, oval or egg-shaped, like the popular Brown Turkey (also known as Common Purple, Brown Naples, Blue Burgundy and Lee's Perpetual). Another popular variety is the violet-coloured Celeste, whose light pink or rose-coloured flesh is particularly juicy and sweet. Celeste figs are grown widely in the south-western areas of America. Also popular are the Brunswick (pale purplish-green skin and pink-tinged flesh, fairly rich and sweet), St Johns (dwarf variety with white-fleshed, pear-shaped fruits) and White Ischia (small, greeny-white figs with juicy purplish flesh).

Some figs have longer necks than others; still others have a 'ribbed' appearance. There are even some white figs, such

In Lancashire, 'Fig Sue' was customarily eaten on Good Friday. This was a sort of hot soup made with sliced figs, chunks of bread and currants, all simmered in ale and spiced with nutmeg.

as Col di Signora Bianca, which have a thick, yellowish-white skin with a rich, sugary, claret-coloured flesh inside. White Adriatic is another variety, found mainly in America, while one of the most famous black figs is Black Mission, grown in California.

In France the most highly prized fig is the Marseilles fig. This is a medium-sized fruit with yellowish-white skin tinged with gold. Also popular in European cuisine are the Bellona fig (large, purplish-skinned fruit with bright red pulp), the black Bannisotte fig, the white and red figs of Argentuil and the Barbillone fig (purplish-skinned fruit with white pulp). In the south of France, *figues sêches* (dried figs) are still prepared in the traditional fashion. Very ripe autumn fruit is spread out on hurdles in the sun and taken indoors at night. They have to be turned over several times during the drying process and, by the time they are completely dried, they are slightly flattened. Prepared in this way, *figues sêches* will keep for a very long time.

Australia has many species of native figs. They are found mostly in the northern rainforests, with fewer species in the south. Fruit from these vary from pea-size to around 5cm. They are usually edible, but of greatly varying quality. They must be soft and pulpy to be palatable, otherwise, like all figs, they tend to be dry and gritty with poor flavour. At Adventure Bay in 1792, Captain William Bligh is recorded as having planted figs, quinces and pomegranates. Aborigines of the desert regions particularly savour the Desert Fig, *F. platypoda*. It has very appetising fruit which turn a deep, bright red when fully ripe. Like the wild ancestors of the Common fig, native figs are most likely to grow in rock crevices where water and leaf debris have accumulated. The aborigines used the rough leaves of the Sandpaper Fig (*F. coronata*) to smooth the handles of their spears and boomerangs, and, similarly, leaves of the *F. serrata*, or Old World Fig, were used for removing rust from iron implements. The fruit also makes good jam.

More commonly seen in Australian gardens and parks is the Moreton Bay Fig (*F. macrophylla*), an evergreen native to New South Wales and Queensland. It can reach 30 metres

'...that God-given inheritance of our mother country, darling of my heart, a dried fig...'
Alexis of Thuril, 4th century BC

in height, with a six metre spread. It is drought-resistant and prefers a medium soil in an open, sunny position. The thick trunk is smooth and grey, while the dark green glossy leaves are smooth and leathery, to about 25cm long. When ripe, in summer, the small purplish fruit may be picked. There are some magnificent specimens of Moreton Bay figs in and around Sydney; however, if considering one for the home garden, be sure to allow ample space for its enormous dimensions. It can be grown from cuttings taken in autumn.

All cultivated figs are sweet when ripe (figs, in general, have the highest sugar content of any fruit), but some varieties are sweeter than others. The fig's 'sweetness and good fruit' (Judges ix:2) and its beautiful dark green foliage have been frequently alluded to in ancient texts. It is thought that Jesus Christ ate figs along the road to Bethany which is why 'Fig Sunday' is celebrated during Lent in England. Of course, one of the better known references to fig leaves is found in Genesis, in the account of the fall of Adam and Eve, for they are said to have sewn fig leaves together to form aprons to cover their nakedness. There is also a reference to the fig as a symbol of peace and harmony:

> But they shall sit every man under his vine and under his fig tree and none shall make them afraid.
> (*Micah 4:4*)

In fact, some writers have suggested that the unspecified fruit that Eve offered Adam was actually a fig, not an apple. All the Bible says is that the fruit grew on the Tree of Knowledge of Good and Evil. Imagine that the fruit was indeed a fig. It does seem a more suitable contender: its leaves were apparently at hand, offering a shield for newly bashful eyes and it also has significance in other ancient religions. As to its powers of seduction, the purple skin may not be as shiny and inviting as the apple's rosiness, but its luscious fruit is far more exotic and suggestive of temptation than the apple's crisp white flesh.

As well as being used for food, figs were used medicinally. The Bible mentions that King Hezekiah was

'Philemon, a Comick Poet, died with extreme laughter at the conceit of seeing an asse eat figs...'
Thomas Nashe
The Unfortunate Traveller (1594)

cured of a boil after his servant followed Isiah's advice and placed 'a cake of figs' on it! The Muslims were also fond of figs and Mohammed is said to have written: "If I desired a fruit in the Garden of Eden, I would choose the fig."

The fig is thought to have originated in Persia, Asia Minor and Syria and it continues to be an important dietary staple in these areas. It grows best in the northern,warm temperate and sub-tropical regions. The Greeks are thought to have obtained their figs from Caria in Asia Minor, which explains the derivation of their Latin name. Figs were singularly well suited to the rather meagre soil of Greece which was unable to bear very good grain crops. They were also known in Crete from very early times; remnants of figs were found in the building identified as the palace storehouse at Minoan Hafia Triada and figs also appear in many of the famed wall paintings of Knossos.

The Greek writer Theophrastus was among the first to write of cultivated and wild forms of figs, along with apples, olives, pears and vine crops. He wrote specifically:

> As for the fig, no cultivated kind is raised from its seed, but either the ordinary wild fig or some wild kind is the result, and this often differs in colour from the parent; a black fig gives a white and conversely...

The Greek geographer Strabo added that the Smyrna fig was highly esteemed and also asked the highest price in the market. Plato was said to have owned extensive fig orchards in Attica and to have described himself as a *philosykos*, or 'fig lover'. The Greeks believed figs were a special gift from the gods to ensure their people would not go hungry and they planted groves of sacred fig trees around Athens to show their gratitude.

Figs were much celebrated in Greek culture and cuisine and they featured widely in the Greek diet. Athletes, for instance, were fed almost entirely on figs for they were thought to bestow strength and swiftness; similarly, the vigorous soldiers of Sparta were thought to have favoured eating figs before a battle. At one time, a sort of sweet cake made by mashing up figs and drying them in the sun was

'Fygges doth nowrysshe more than any other frute...They do clense the brest and the lunges... and also they doth provoke a man to sweate; wherefore they doth engender lyce...'
Andrew Borde (1547)

popular throughout the Greek archipelago as a substitute for bread. In fact, the figs of Attica were so important to the people of ancient Greece that a law was passed which prevented the exportation of the best quality fruit.

The fig spread from the eastern Mediterranean in other directions, wherever trade routes were opened. The Phoenician trade caravans took dried figs along 'the silk route' in India and China. (The fresh ones, of course, did not travel well, owing to their fine skin). As early as the third century BC, an Indian ruler asked the Greeks to send him 'figs, grape syrup and a philosopher'. They filled the first two items of the order, but had to explain that philosophers were not an item of commerce! The Arabs went on to extend the fig's range westward from Egypt in the eighth and ninth centuries.

Figs were introduced to Italy with great success. Fossilised remains of dried figs have been excavated in Pompeii and, along with many other fruits, figs and fig trees are featured in the murals and wall-paintings of that once-buried city. Recent excavations in Pompeii have shown the existence of orchards which supplied figs to local shops, along with other small fruits. The orchards on the slopes of Vesuvius were famous for the quality of their figs, some of which were exported to England—before the eruption of the volcano, of course—to supply the Roman forces which were stationed there.

The Romans were adept in the arts of grafting and budding, making use of the best selected varieties of fruit trees which were worked onto rootstocks, usually grown from cuttings. Columella, who lived in Italy in the 1st century AD, was a noted Roman authority on the culture and uses for many fruits and he devoted many pages of his writings to this subject. Much of his advice about figs, quinces, grapes and medlars is very sound and could be adapted today with little modification. Columella also wrote, rather lyrically, about the different varieties of figs:

> But 'tis beneath baleful Arcturus' star
> That Livian figs, which with the Chalcidian vie.

Sugared figs, pressed into phallic shapes, were once sold by peddlers in Kyoto as part of Japan's Shinto Autumn Festival.

> And Caunion, rivals of the Chian, come
> To birth, and purple Chelidonian
> And fat Mariscan, Callistruthian
> Gay with rosy seeds, and that pale kind
> Which ever keeps the name of tawny wax
> And the cleft Libyan and the Lydian fig
> With spotted skin.

The Roman Varro (116-27 BC) wrote extensively of wine production, along with figs, mulberries and the propagation of pears and apples. He also made an interesting comment about how to successfully transplant figs out of the country:

> It is better to plant the nursery shoots from the fig trees than from the seeds of the fruit: unless this is impracticable, as when you wish to ship overseas or import from thence. In this case we pass a string through the figs when they are ripe for eating, and after they have dried they are tied in bundles and may be sent where we will: and these are planted in a nursery and reproduce. It was in this manner that the Chian, Chalcidian, Lydian, African and other varieties of overseas figs were imported into Italy.

Cato also gave advice on the different varieties of figs and how best to cultivate them:

> Plant Mariscan figs in chalky open soil; the African, Herculanean, Saguntine, the winter variety and the black Tellanain with long pedicels, in soil which is richer or manured.

Legend has it that the Romans continued to rate imported foreign figs above their own domestic varieties. They imported them from the Greek island of Chios and from Carthage. Pliny gives details of no less than 29 kinds that he was aware of, giving special praise to those from "Tarant and Caria and also those of Herculaneum". Pliny went on to write at length about figs and his comments indicate the high regard Romans had for this fruit:

> The figs that are highly approved are given the distinction of being dried and kept in boxes, the best and largest growing on the island of Iviza and the next best in the district of Chieti; but in some places where there is a very large supply of them, they are packed for storage in large jars in Asia, but in casks in the city of Rusina in Africa and when dry they serve the purpose of bread and other viands at the same time, inasmuch as Cato, as if laying down a law as to the proper rations for agricultural labourers, prescribes that they are to be reduced in quantity during the time when the figs are ripe.
>
> A plan has been derived to use a fresh fig instead of salt when eating cheese. To this class, as we have said, belong the Syrian and Carian figs and the Caunean figs that, when Marcus Crassus was embarking to sail against the Parthians, gave him an omen by the voice of a man crying them for sale.
>
> All these varieties of fruit were imported from Syria to his country place at Alba by Lucius Vitellius, afterwards censor, when he was Lietenant-governor of that province, in the latter part of the principate of the emperor Tiberius...

Excavation of the early Roman fortresses has revealed the remains of figs and they were also found on what is thought to be the site of an army hospital, suggesting they had been prescribed as nourishing food for recuperating soldiers.

Pliny wrote that figs are '...the best food that can be taken by those who are brought low by illness and long sickness and are on the way to recovery.' He also stated that figs, especially in the fresh state, formed a large portion of the fare of slaves and agricultural workers. He went on to describe a sort of wine, known as *sycites*, which was made from ripe and fermented figs. Interestingly, a fig wine or *eau-de-vie* called *boukha* or *mahia* is still to be found in certain North African countries, notably Tunisia. There it is served

Cleopatra's favourite fruit was the fig and the asp with which she killed herself was brought to her in a fig basket.

both before a meal as an aperitif and afterwards, with coffee, as a liqueur. The Germans also favour a fig brandy which goes by the name of *Feigenbranntwein*.

It is amusing to learn, given the rather lush and corrupt nature of later Roman society, that the wealthy Apicius fed expensive, imported Smyrna figs to his hogs to improve the flavour of their meat! Less exalted breeders used dried domestic figs and honey to give their poor pigs hypertrophied livers, from which they made pork *foie gras*. This was the first step towards what we know as *foie gras* today.

Many Greek legends sprouted around the fig. In one, Sykeus and his mother, Ge, were pursued by Zeus during the war of the Titans. One of Zeus' lightning bolts struck Sykeus and Ge metamorphosed her son into a fig tree so he could continue to live. The myth persists that the fig tree can avert lightning strikes.

An Athenian legend attributes the revelation of the 'fruit of autumn' to mortals to the goddess Demeter who rewarded King Phytallus for receiving her into his home. A fig tree stood by Phytallus' grave, and the whole area was referred to as Holy Fig Tree Hill. The Greeks held the plant and fruit in great reverence and believed that a priestess could tame a wild bull and lead it to the sacrificial altar simply by tying a branch of wild figs around its neck.

The word 'sycophant'—from *sykos* 'fig' and *phanein* 'to show'—possibly originates from a time of scarcity of figs in Attica. Figs were stolen from trees sacred to the gods and, later, certain people were accused and were found to have the figs in their homes. Samuel Taylor Coleridge wrote that the word sycophant meant 'a wretch who flatters the prevailing party by informing against his neighbours the pretence that they are exporters of prohibited figs or fancies.'

Figs played an important part in Roman mythology, too. They were dedicated to Bacchus and featured in many lusty ceremonies in that god's honour. The wolf that suckled Romulus and Remus was said to have slept near a certain fig tree which was subsequently considered sacred by the people of Rome. Pliny ascribed the mystic prowess of this sacred tree thus:

> A fig tree growing in the actual forum and meeting place of Rome is worshipped as sacred because things struck by lightning are buried there and still more as a memorial of the fig tree under which the nurse of Romulus and Remus first sheltered those founders of the empire on the Lupercal Hill. The tree that has been given the name of Ruminalis, because it was beneath it that the wolf was discovered giving her rumis to the infants. And it is also a portent of some future event when it withers away and then by the good offices of some of the priests is replanted. There was also a fig tree in front of the temple of Saturn which in 404 BC, after a sacrifice had been offered by the Vestal Virgins, was removed, because it was upsetting a statue of Silvanus.

The philosopher and historian Ovid tells us that figs were exchanged as presents on the first day of the new year, to symbolise abundance and luck for the future. The inhabitants of Cyrene wore wreaths of fig leaves when making sacrifices to Saturn, for he was believed to have first discovered the fruit.

Interestingly, the early Christians overlooked the fig's pagan connotations and revered it also. An old legend tells us that on her journey into Egypt the Virgin Mary hid herself and the Infant Jesus in a fig tree to escape Herod's fury. When the Virgin passed, the fig tree opened to receive her, then closed its branches tightly to protect her; when Herod's soldiers had passed, the fig opened its branches again and the Virgin was able to proceed on her journey. This particular fig tree still stands at Mathave, where it is often pointed out to tourists.

Another early Christian tale regarding figs was recorded by Matthew (21:19) when he was observing the entry of Jesus into Jerusalem: 'When he saw a fig tree in the way, he came to it, and found nothing thereon, but leaves only, and said unto it, Let no fruit grow on thee henceforward ever. And presently the fig tree withered away.'

The Romans introduced figs to other European countries

A Garden Song
All the seasons run their race
In this quiet resting-place;
Peach and apricot and fig
Here will ripen and grow big;
Here is store and overplus—
More had not Alcinous!
Austin Dobson (1840-1921)
(quoted in *The Golden Treasury*,
J. M. Dent & Sons, London,
1906)

and are thought to have also brought them to England, or southern Gaul as it was then known. Specifically the variety known as Reculver was thought to have been named after Reculver in Kent, an important Roman port. However, it is likely that after the Roman occupation figs disappeared, to be reintroduced several centuries later in different varieties, when there was a revival of interest in gardening, led by the monasteries. An account given by a 14th century gardener named John Harrison attests to the great improvement in knowledge of '...such herbes, fruites and rootes which grow yearlie out of the ground...but which (had previously) remained either unknowne or supposed as food more meet for hogs and savage beasts to feed than mankind.' He went on to describe the positive influence of the teaching of horticultural methods by the monasteries which resulted in all sorts of fruit and vegetables becoming available to both the rich and the poor: '...so we have no lesse store of strange fruites as abricots, almonds, peaches, figges and conetrees ...as in noble men's orchards.'

Dried figs appeared in Paris in the 14th century, where they were served both as h'ors d'oeuvres and desserts. They were imported in large quantities from Italy and Spain to cope with the demand. Dusted with powdered sugar, figs reached new heights of popularity during Tudor times as a 'kissing confect' or lovers' sweetmeat. Also during this period, stews and venison were often improved by the addition of dried or preserved figs and spices to mask the taste of meat that was usually far from fresh. In 1575 Mascall detailed the propagation of the fig, writing:

> The black fygges are the best, being dryed in the Sunne, and then layd in a vessell in beddes one by another. If a fygge will not beare, ye shall dygge him all about, and under the rootes in Februarie, and take out all the earth and put unto him the doing of a privie, for that he lyketh best...

Figs are . . . perhaps the most sensual fruits, with their blooming, bursting skins and lush flesh.

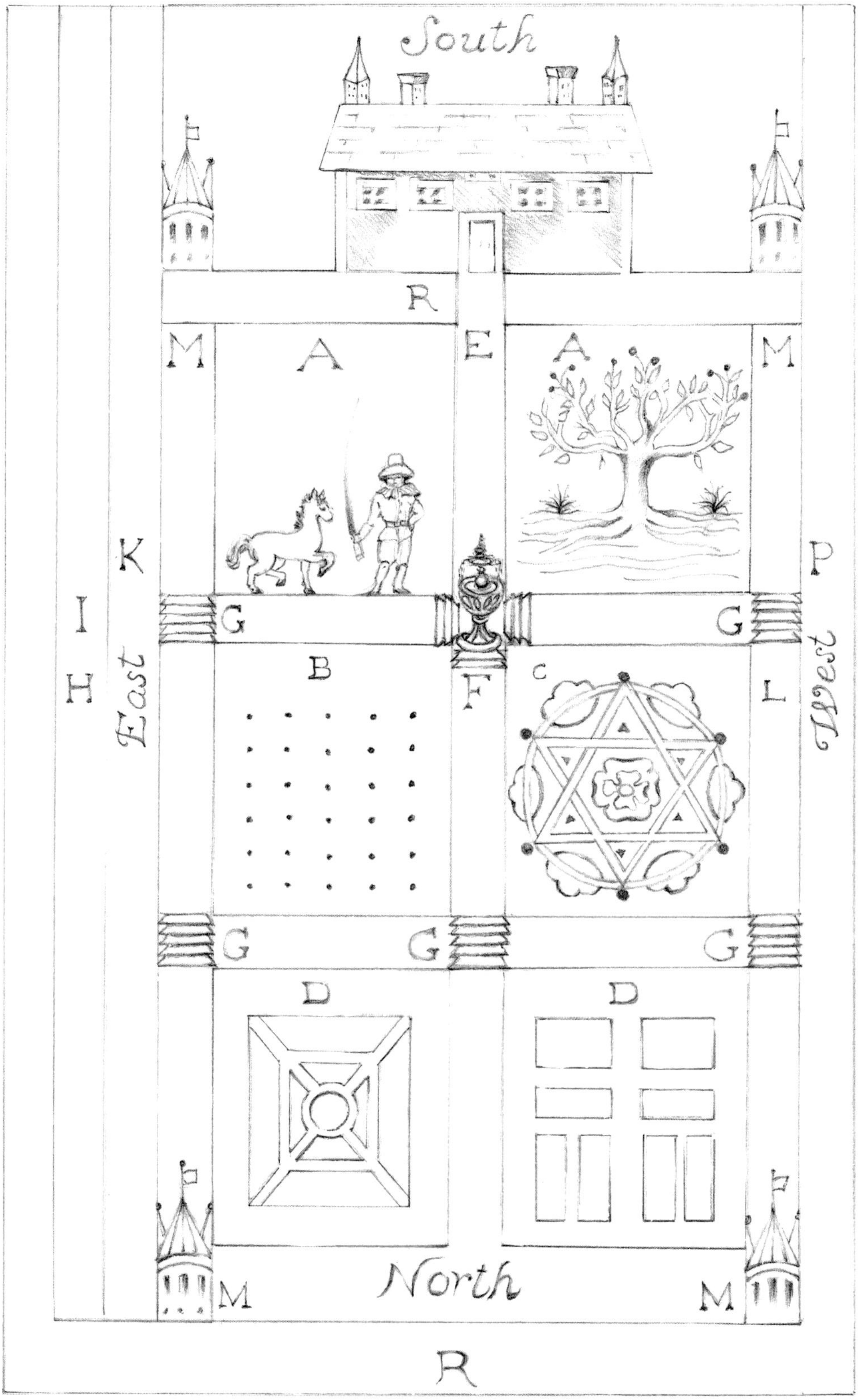

Outside the courtyard, near the entrance gates, is a garden of four acres with a fence running around on either side. Here grow tall flourishing fruit trees: pears, pomegranates and shining apples, sweet figs and luxuriant olives... Pear after pear comes to maturity and apple after apple, one bunch of grapes after another and fig after fig...

By 1597, herbalist and noted gardener John Gerard was to write:

> The fig trees doe growe plentifully in Spaine and Italie, and many other countries, as in England, where they beare fruite but it never commeth to kindly maturities, except the tree be planted under a hot wall, whereto neither north nor north east winds can come.
>
> The dwarffe fygge groweth in my garden and bringeth foorth ripe and very great fruit in the moneth of August, of which Figs sundrie persons have eaten with great pleasure.

Given its growing popularity, it is somewhat surprising that John Parkinson, in 1629, was only able to list three varieties of figs. They were, however, widely planted:

> The figge trees that are noursed up in our country are of three sortes. whereof two are high, the one bearing against a wall, goodeley sweeter and delicate Figs called Figs of Albania, and is blewish when ripe; the other kind tall is nothing so goode and may be the white ordinary kind from Spaine. The third is a dwarf kind of Figge tree, not growing higher than a mans body or shoulders, bearing excellent good figges and blew, but not so large as the first kind.
>
> Figges are served to the table with Raysins of the Sunne and blanched Almonds for a Lenten dish. The Figs that growe with us when they be ripe and fresh gathered, are eaten of a divers with a little salt and pepper as a dainty banquet to entertain a friend, which seldom passeth without a glasse of wine to wash down.

In ancient Greece athletes were given dried figs to eat. This was thought to confer extra strength and stamina.

Several varieties were to be introduced later in the century and their numbers were also to be greatly augmented by imports, especially through the activities of John Tradescant, who travelled extensively abroad collecting plants from Russia, North Africa and parts of America. It was during this time that figs became commonly grown against walls for walls were increasingly built to surround the gardens of the large houses and mansions which had replaced the moated castles of previous unsettled ages. At this time, surplus fruits and vegetables were sold directly from these large London gardens of the wealthy and it is likely that they included figs, along with peaches, quinces and apples.

William Shakespeare rather inconsistently had Iago exclaim contemptuously: 'Virtue! A fig!' but let Cleopatra's assistant Charmian sigh plaintively: 'I love life better than figs.' French author Georges Blond sided with Charmian when he described the fig as '...the manna of the Mediterranean.' By the time of Elizabeth I, every important English household kept dried figs on hand to make puddings. Germans started eating figs at about the same time. Northern France acquired fig trees when they were planted in the gardens of Louis XIV, who delighted in all new and exotic foods. In fact, during the reign of 'the Sun King', the cultivation of figs became somewhat of a craze amongst the decadent French aristocracy. By the latter part of his reign the nurseries attached to the Royal gardens were said to have contained many hundreds of varieties of all sorts of fruits.

Interestingly, the fig was known to Americans very early in the piece. Spanish colonisers introduced it to the New World, a fact easily guessed from the name of one of California's most successful varieties—the Mission Fig. Figs were established in Florida before the end of the 16th century and, in 1629, John Smith recorded that a certain Mistress Pearce, a settler at Jamestown, Virginia, harvested 'neere a hundred bushels of excellent figges.' Today California produces over 90 per cent of all figs grown for commercial use in the US.

By 1708, according to Henry Wise, gardener to Queen Anne, eight varieties of imported figs grew in the gardens of England's Hampton Court, Windsor and Blenheim. Loudon, writing in 1834, described a Sussex orchard as having '...one hundred trees, each yielding every day a dozen figs the size of apples for sale in nearby watering places.' He may have been referring to the fig garden at

The Italians say fig leaves are unlucky and believe that evil spirits lurk in them during the summer months.

Bishop's Garth in West Sussex, which dates from 1745.

Today figs are cultivated in sunshine and fertile soil, just as they were over 3000 years ago in the Fertile Crescent, an area stretching from Persia to Egypt. The gnarled trees particularly enjoy warm Mediterranean conditions and they have recently been planted on a commercial scale in Madagascar, Australia and South Africa. Figs are grown commercially in California, Portugal, Turkey, Algeria, Italy and Greece.

Figs will not stand frost and require some protection in winter. The best position for a fig tree is against a sunny wall in a paved, walled courtyard. Not only will the wall afford protection from the cold, but, by confining the roots within the walled courtyard, any tendency to over-luxuriant growth is curtailed. Also, as the fig tends to spread out rather than reach up, it is easy to train it against a wall. From the urban gardener's point of view, such a fan-trained fig is a delightful ornamental feature on a patio or in a courtyard. Confining the plant's roots is actually advantageous because, when they are restricted, the tree will bear a better crop of fruit than if they were allowed to roam. (By the same token, large pots or tubs can be used.) In fact some gardening texts even recommend the very old method of constructing a slightly raised brick or tile border around the fig tree,or confining it with a brick or concrete path which flanks the wall against which the fig is planted.

In Australia, the heavy fruit crop is usually attacked by birds, such as the native silvereye and the imported bulbul so some extra protection, such as netting, is required at fruiting time. Fig trees should be protected from wasps, too, which attack the fruits just as they begin to swell and colour up. Young shoots, especially if overcrowded, can get a grey mould on them in very wet seasons and this, or any sign of canker, is best pruned away when first noticed.

Ficus carica is an attractive, small, deciduous tree or shrub with palmate leaves of usually three or sometimes five lobes. These are rough-textured on the top side and slightly downy on the underneath. It will rarely exceed 30ft (10m). In fact, cultivars are usually less vigorous again and, grown

An old European saying warns that if you eat figs out of season you will be very unlucky.

as a standard or trained against a wall, they will require very little pruning. Regular 'pinching out' of new, young shoots in midsummer will produce plenty of good strong growth. Remember that figs only mature on new wood so it should be your aim to ensure there is plenty of this. To encourage a good supply of frost-resistant, stubby wood that will carry the embryo figs through the winter, the growing points of the new shoots should be pruned in mid summer to about five leaves. Always cut away any weak or deformed branches, taking the growth back to the trunk, if necessary.

Apart from these special requirements,the fig is a sturdy and attractive tree and will practically look after itself. In fact, the trees will live to a great age, as evidenced by the antique specimens said to have been planted in the garden of the Archbishop's Palace at Lambeth in 1525. By 1548 horticulturist William Turner was able to note in his *The names of herbes* that there were '...divers fig trees in England in gardens' and, according to writer John Loudon, in 1834 these same trees still grew in the Lambeth Gardens, having attained a height of 50 feet, despite many storms and severe winters. Even older is the fig tree which grows in Anuradhapura, north of Kandy, Sri Lanka, which is said to have come from a cutting of the Bo tree (*Ficus religiosa*) under which Buddha once sat. It is said to have been sent to Sri Lanka in 288 BC, as a gift from the Indian Emperor, Asoka.

Other key elements for the successful cultivation of figs include a well-drained, slightly chalky soil and a bright sunlit aspect which will help ripen the fruit to perfection. Do not plant fig trees too close to a vegetable patch—the nitrogen-based fertiliser used to feed vegetables will have a negative effect on the fig. For the best flavour, sprinkle powdered lime into the soil around the fig tree. Similarly, avoid using manure or compost around the base of a fig, as these are both rich sources of nitrogen. The only time when the use of manure is warranted is when the tree is bearing heavily. The tree can also be given a little bonemeal each spring if you fear the soil nutrients are being leached out. Do not stint your fig with water during its growing season; like all root-restricted plants, they need more water than those grown in regular borders and generous watering will

They sewed fig-leaves together and made themselves aprons.
Old Testament—Genesis iii:7

help to produce juicy fruit.

Figs are ready to be picked when a drop of nectar collects in the 'eye' opposite the stem end and the fruit feels quite soft when gently picked. If the figs are not to be eaten at once, store them in the bottom of the refrigerator.

Considered botanically, the fig is a unique and remarkable thing, being neither fruit, flower, nor seed. It is, instead, a hollow, globular or pear-shaped receptacle enclosing hundreds of tiny male, female and neuter flowers which come to full perfection as the 'fruit' ripens, creating a mass of tiny, one-seeded fruits which grow and mature without ever seeing daylight. Interestingly, this is quite the reverse of the strawberry which is, in fact, a fleshy receptacle with many seeds dotted over its exterior.

Figs are produced singly or in pairs at the junction of the leaf axils. The trees tend to produce two crops of fruit each season. The first one is quite often a failure, with the figs dropping before they are ripe and this is most often due to the inadequate fertilisation of the female flowers by the very few, smaller male ones. Some cultivated varieties or hybrids produce no male flowers at all and this is where Mother Nature has stepped in to lend a helping hand in a somewhat bizarre way. It has been found that successful fertilisation is often facilitated naturally by the entry of insects through the very small orifice which remains open in the flowering fig. At least as long ago as the time of Aristotle, it was known that the fig-bearing female tree was fertilised by a fig wasp, *Blastophaga psenes*, which transferred pollen from the male fig flowers to the female's by crawling inside each fruit. Pliny was later to describe the artificial fertilisation process as follows:

> A plan was discovered of bringing branches of the wild fig, the Capri, from somewhere else and throwing them tied in bundles, on the fig orchard. A treatment which orchard figs do not require when planted on thin soil with a northerly aspect, since they dry of their own accord owing to the situation of the place, and this cause, by making them split open, produces the same result as the action of the wasps.

The last garden of my girlhood was at Glenmore, the home of my cousin, the late Mr James Riley. He, like myself, was passionately fond of flowers, making periodical additions to his already fine collection. Not long since I last saw the vestige of my 'lost bower' —a fence post, covered with ivy-growing fig (*Ficus stipulata*) now become nearly a tree, and, what is extremely rare, bearing its fruit.

Mrs Rolf Boldrewood, *A Flower Gardener's Calendar* (1893) from *The Flower Garden in Australia, a Book for Ladies and Amateurs* (quotes in *The Australian Gardener's Anthology*)

When Smyrna figs were introduced to California at the end of the 19th century, fig wasps had to be introduced, too. The promulgation of this information by the US Department of Agriculture in 1900 saved the then fledgling fig industry.

In earlier times, patient fig growers would adopt another artificial means of ensuring fertilisation by inserting a small feather in the cavity and twirling it around, thus ensuring the pollen brushed against the pistils. This process was called 'caprification' from the Latin for 'wild fig', *caprificus*, referring to the custom in countries where figs grew wild of shaking branches of flowering wild fig over the cultivated trees so the pollen would ensure the development of the fruit. For the sake of the urban gardener who just wants to grow a fig tree or two, however, it is worth pointing out that many hybrid figs are available today and these set fruit without the aid of pollination.

Fresh and dried figs are both delicious and nutritious. Both provide a good energy source along with moderate fibre and some vitamin C. Figs are unusual in that they also supply magnesium, calcium and iron. Interestingly, figs' food value improves with drying, in much the same way as that of dates. Most packeted, dried, canned or glacé figs come from Asia Minor, Spain and the south of France. Natural sun-dried figs are loose-packed and retain, to some degree, their original shape. Plumped or 'pulled' figs have usually been kneaded to make them supple and easier to pack attractively. Bay leaves are often packed in each box to keep the fruit free from mould and to keep destructive insects at bay. 'Pressed' figs are so-named because they are so closely packed into boxes that they become compressed into discs.

All thy strongholds shall be like fig trees with the first ripe figs; if they be shaken, they shall even fall into the mouth of the eater.

Old Testament: Nakum iii:12

Whether you choose to eat them dried or fresh, figs may be successfully combined with many other ingredients and have possibilities limited only by the imagination. In Italy and Greece, figs are traditionally served as an *antipasti* with slices of meat such as fresh prosciutto, or on their own with freshly ground black pepper. Ginger, cream, nuts, cinnamon, crème fraîche and citrus fruits all go well with figs. So do vanilla, vinegar, red and white wine, sherry, port, juniper, fennel, cloves, mustard, coriander and allspice. Coconut cream is a delicious ally of the fig as well.

Serve fresh figs as a dessert fruit, on cheese boards and fruit platters, stuffed with either a sweet or spicy filling, or added to ice cream, sorbets or cakes. Sauté fresh figs in a dot of butter and a little brown sugar, flame them with brandy and then serve with double cream for a luscious and quick dessert. Eat them grilled and served as a dessert or as an accompaniment to meats; fill them with various mixtures (cream cheese and blue cheese are especially delectable), or toss them in fresh fruit salads. Dried figs may be chopped or minced and added to puddings, breads, cakes, mousses and ice cream. Deliciously different combinations include sausages, nuts, buttermilk, sour cream and pork.

Stew dried figs and serve them for breakfast or for dessert; or, macerate them in port and serve with ricotta cheese. Use dried figs in any recipe calling for dried apricots —delicious every time! Liqueurs, especially maraschino cherry liqueur, are another interesting addition to this sugary syrupy, uniquely flavoured fruit. Figs in port are particularly heady. Place the sliced figs in a deep bowl and just cover with port before chilling for an hour or two. Then top with lightly whipped cream sweetened with a dash of honey and perhaps a squeeze of lemon juice. Or, use warmed apple or grape juice instead of a liqueur; pour over the figs with a handful of sultanas and chill. Delicious!

And, don't forget the wonderful jam-, sauce- or preserve-making opportunities afforded by figs, not to mention surprising ways they can be used to complement meat and poultry. This book offers a wide selection of recipes, from tried-and-true easy puddings, jams, cakes and desserts to more unusual—and sometimes spectacular—dishes from around the world and from days gone by.

...a few drops of either Sherry or Marsala inserted with an eyedropper into the fig before serving will win much applause for the gardener.

Edward Bunyan, *The Anatomy of Dessert* (1933)

In early times, farmers and gardeners alike would be sure to bury the first fig of the season under a fig tree. This was thought to make the tree bear well and probably harks back to primitive customs where sacrifices were made to the trees to propitiate the gods who presided over them.

Kitchen Quips

- ☆ To store dried figs, place in an airtight jar in a cool cupboard. Put a strip of fresh orange peel (which should be renewed from time to time) in the container with the fruit. This will help maintain freshness and quality. Dried figs can also be stored in the freezer for long periods and they will thaw quite quickly at room temperature.
- ☆ If dried figs have not been tenderised, lightly oil your knife or kitchen scissors to make them easier to cut.
- ☆ If a recipe calls for dried figs to be plumped up before use, experiment with strained black coffee, fruit juices, wines or other liquids for different flavours. Simply cover the figs with the liquid and allow to stand for several hours or overnight (depending on the degree of softness required).
- ☆ Figs do not contain a great deal of natural fruit acid so, if you wish to preserve them whole or as part of a dish, ensure that either citric acid or lemon or orange juice is included in the recipe.
- ☆ When a recipe calls for whole figs—such as preserving the fruit in syrup, or poaching them in liquor—be sure to cut well into the flesh when removing the stem. This will ensure the preserving liquid thoroughly penetrates the whole fruit.
- ☆ To freeze figs, best results are obtained when sweet, soft, almost overripe fruit is used. Wash, remove stems and peel, if desired. Crush, slice or use whole. For whole fruits, cover with a light sugar syrup. Whether freezing the flesh or the whole fruit, lemon juice should also be added to the mix; the ascorbic acid will reduce discolouration.
- ☆ Don't buy figs that give off a sour odour. This indicates overripeness. Instead, look for softly ripe, sweet-smelling fruit.
- ☆ If you are not going to use fresh figs immediately, store them in the refrigerator. As they are extremely perishable, try to use them within one to two days.

Illicit figs

Although this is law-breaking information, it is given here as a matter of interest.

Famous throughout Morocco and Tangiers is *madjoon,* colloquially known amongst Westerners who have sampled it as 'hashish fudge'. Indeed, it is cut and served in little squares and does have a chocolatey taste, just like fudge—though it has a far more potent effect! Some types are compounded of many different spices and can be bought from shops by the kilo, measured out just like sugar. A 'secret ingredient' is said to be ground porcupine needles... Chopped dried figs, seedless raisins, aniseed, cloves, crushed walnuts and honey are also traditional ingredients found in this truly intoxicating dish.

Just Desserts

Enjoy these delicious, irresistible desserts made from figs, combined with fruit, nuts, chocolate and other delicacies. Some recipes are based on old-fashioned favourites, others are delectable and new ways to provide fabulous finales to a special dinner.

During the Roman festival of Saturnalia, statues of the gods were crowned with garlands of fig leaves and fresh figs were piled at their feet to symbolise a plentiful harvest and fortunate times for the year to come.

FIG AND GINGER PUDDING

110g/4 oz butter
200ml/7 fl oz honey
25ml/1 fl oz golden syrup
4 eggs
130g/5 oz flour
1 tspn baking powder
200g/7 oz breadcrumbs
100g/4 oz glacé figs, chopped
85g/3 oz preserved or glacé ginger, chopped
285g/8 oz candied lemon peel
75g/2½ oz walnuts, finely chopped
½ tspn each ground ginger, nutmeg and cinnamon
75ml/2½ fl oz cup milk
whipped cream and extra candied lemon peel, to garnish

Grease a pudding mould with vegetable oil. Cream butter, honey and golden syrup together, then mix in eggs, one at a time. Sift flour and baking powder together and add to butter mixture, then stir in the rest of the ingredients and mix till smooth. Pour mixture into mould and cover tightly with absorbent paper and then with a layer of foil.
Place the bowl in a roasting tray and place on rack in moderate oven 180°C (350°F). Pour boiling water into the tray till it comes halfway up the outside of the mould. Steam for 4 hours, adding boiling water as necessary to maintain level. Pudding is cooked when top springs back to a light touch. Allow to cool, then run a knife dipped in hot water around the edge of mould before inverting onto serving platter. Serve with whipped cream, garnished with chopped or finely grated candied lemon peel.

FIG AND COINTREAU RING

1 cup chopped dried figs
1 cup red-skinned apple, diced
1 cup tinned mandarin segments, drained and diced
¼ cup tinned grapefruit segments, drained and diced
1½ cups reserved mandarin syrup (made up to 1½-2 cups

By the superstitious, the fig tree was viewed with much respect and some reservation—after all, some believe it was the tree which first gave knowledge of evil:

> And the eyes of them (Adam and Eve) were opened and they knew they were naked; and they sewed fig leaves together and made themselves aprons...
>
> (*Genesis* iii 7:8)

It was also thought that the devil visited fig trees and lurked there, waiting for children who came to eat the fruit.

with extra hot water)
¼ cup sugar
2 tbspns gelatine
1½ cups Cointreau liqueur
1 tspn orange zest
2 tbspns orange juice
few drops orange food colouring, if desired

Prepare gelatine with the sugar and mandarin syrup/water mixture, according to manufacturer's instructions. Add orange juice, orange zest, Cointreau and food coloring; chill until syrupy. Combine figs, apple, mandarin and grapefruit segments in bowl, then stir carefully into jelly mixture. Pour into ring mould and leave until set. Unmould by dipping into hot water for a few seconds before inverting onto a chilled glass platter.

FIG AND HAZELNUT CREAM

2 cups milk
½ cup cream
1 tbspn caster sugar
1 tbspn cornflour
30g/1 oz butter, softened
1 tbspn hazelnuts, finely chopped
2 cups chopped dried figs
whipped cream flavoured with a dash of almond essence, to serve

Combine milk, cream and sugar in double saucepan over medium heat and bring slowly just to the boil; reduce heat and continue to simmer. Mix cornflour and butter together, adding a little milk to thin mixture slightly, then stir cornflour mixture into milk; continue to stir over heat for 2-3 minutes, but do not allow to boil. Stir through figs and hazelnuts and then pour mixture into a wetted ring mould and refrigerate for 2-3 hours.

To turn out, run a knife, which has been dipped in hot water, around the edge of the mould, then tip out onto chilled plate and garnish with rosette of whipped cream. If desired,

Figs are soft, easily digested and a good source of energy. They were commonly eaten by Roman gladiators, which is surely indicative of their strength-giving qualities.

On Christmas Eve in Provence, to occupy the time before going to Midnight Mass, thirteen desserts are put on the table, symbolising Christ and his disciples. Traditionally, one of the dishes will be 'poor man's nougat'—dried figs which have been halved and stuffed with finely chopped walnuts before being pressed back together.

mound fresh berries — such as blueberries or grapes — in the centre, along with two quartered fresh figs.

FRAGRANT FIG AND PEAR CHARLOTTE

1.5kg/3 lb ripe pears, peeled, cored and chopped
1 tspn grated orange zest
1 tspn grated lemon zest
1 tspn vanilla essence
½ cup fig jam
¼ cup orange juice
12 slices brown bread, crusts removed
2 cups vanilla custard
2 cups berries (eg: raspberries, blueberries), to garnish

Combine pears, orange and lemon zest and vanilla essence in saucepan over low heat for about 30 minutes, stirring frequently until pears are very soft. Drain off any excess liquid and reserve mixture. Combine jam and orange juice in a second saucepan over low heat until melted together. Set oven at 230°C/450°F. Prepare an 8½ inch round glass or china baking dish, either by spraying with non-stick spray, or greasing *very* lightly with butter. Place a round of waxed paper in the bottom of the dish. Cut 5 slices of bread in half diagonally and dip one side of each in the jam mixture. Fit slices, dipped-side down, into the dish, trimming the slices as necessary to fit the bottom. Cut remaining slices into strips, dip into jam mixture and fit around sides of baking dish to cover. Brush any remaining jam mixture over bread. Spoon pear mixture into the centre. Bake at 230°C/450°F for 15 minutes; reduce oven temperature to 180°C/350°F and bake for a further 30 minutes. Let cool and refrigerate overnight. To serve, run a knife around the edge of the dish to loosen the charlotte, invert onto a serving dish and serve with custard and berries.

According to biblical references, a 'lump of figs' laid on King Hezekiah's boil brought about his speedy recovery.

FIG AND COINTREAU COMPOTE

1.25kg/2½ lbs fresh purple or green figs
2 cups water

Although Easter is the main holiday in Greece, Christmas is also a period of considerable celebration. On Christmas Eve, before the family goes to Mass, boys and girls gather in groups to sing carols in front of the houses, accompanied by the beating of small drums and the tinkling of metal triangles. Generally they are rewarded with dried figs, walnuts or almonds and, in the towns, money.

I cup orange juice
⅔ cup sugar
I cup Cointreau
few long fine shreds of orange peel
½ tspn almond essence
thick pouring cream, to garnish

Wash and quarter figs and set aside. Combine water, juice and sugar over medium heat, stirring until sugar is dissolved; bring to the boil and cook for a further five minutes to form a light syrup. Add figs, Cointrequ, orange peel and essence and reduce heat; simmer gently for further five minites. Remove figs and set aside. Bring syrup back to the boil and cook rapidly until it is reduced by about half; pour this syrup over figs, cover and chill. Serve Figs with syrup in individual glass compote dishes, with cream to garnish.

FROZEN FIG-PEACH SUNDAE

6 figs, peeled, sliced and frozen
6 peaches, peeled, stoned and frozen
I cup fresh raspberries
½ cup seedless green grapes
½ cup frozen puréed strawberries
coarsely chopped hazelnuts, for topping
few mint leaves, to garnish

Process figs and then peaches in blender until you have two separate sorbet-like mixtures. Place scoops of fig mixture and peach mixture into sundae glasses, followed by a layer of raspberries, then grapes, then raspberries again. Top each sundae with scoops of frozen puréed strawberries, sprinkle with hazelnuts and garnish with mint.

Tip: Use a tiny melon baller rather than a standard ice cream scoop for a very dainty effect.

For hoarseness — boil well five figs, and five dates, and licorice root, and let the patient eat them.
The Housekeeper's Receipt Book (1816)

PEPPERED FIGS

12 ripe purple figs
juice of I lemon

In this climate, one can eat with safety almost any quantity of fruit.
'Feed him with apricocks and
dewberries,
With purple grapes, green figs
and mulberries.'
Joseph Phipps Townsend
A New South Wales Garden
(1848)
from *Rambles and Observations in New South Wales with Sketches of Men and Manners, Notices of the Aborigines, Glimpses of Scenery and Some Hints to Emigrants* (quoted in An Australian Gardener's Anthology)

500ml/16 fl oz red wine
1 tspn each freshly ground black pepper, cinnamon and nutmeg
brown sugar, to taste

Peel and slice figs and place in bowl with lemon juice. Combine wine and spices, pepper and sugar to taste in a non-aluminium saucepan and bring to the boil. Pour over sliced figs, cover and chill for at least 48 hours before eating. To serve, fill individual bowls with fig and wine mixture and garnish with whipped cream or crème fraîche.

FIG AND PLUM BRULÉE

250g/8 oz fresh plums, stoned, skinned and diced
100g/3½ oz dried figs, chopped
3 tbspns caster sugar
3 tbspns water
4 tbspns kirsch
2 level tspns arrowroot
extra 1 tbspn cold water
¾ cup sour cream
¾ cup double cream
brown sugar
ground cinnamon

Combine plums, figs, caster sugar, kirsch and 3 tbspns water over low heat and simmer for 10 minutes, or until plums are mushy and fig pieces plumped. Stir arrowroot into extra tbspn water and blend, then stir into fruit mixture and continue cooking for one minute.

Spoon portions of fig and plum mixture into individual ovenproof dessert plates or ramekins. Whip cream and sour cream together and spread over the top of each dessert. Sprinkle tops with brown sugar and cinnamon and place under preheated griller just long enough for sugar to caramelise. Serve immediately.

Whether approached by the river or the road, the picturesque appearance of Perth cannot fail to excite admiration…I was much struck by the fig-trees in the bishop's garden close to the river; they were of such great size that I mistook them at first for horse-chestnuts.

Mrs Edward Millett
Perth and its Gardens
from *Parsonage or the Settler and the Savage in Western Australia (1872)*
(quoted in An Australian Gardener's Anthology)

FIG AND PASSIONFRUIT COMPOTE

12 large ripe figs, halved or sliced
100g/3½ oz dried apricots, chopped
1 tbspn currants
orange juice and sweet white wine, to cover fruit
1 tbspn finely chopped glacé ginger
1 tbspn orange zest, grated
1 tspn lemon zest, grated
1 tbspn golden syrup
6 passionfruit
natural or vanilla flavoured yoghurt, to garnish

Combine apricots and currants in bowl, pour over enough orange juice and sweet white wine to cover and leave overnight.

Place fruit mixture in non-aluminium saucepan with ginger, zests and golden syrup, cover and simmer over very low heat for 30-40 minutes. Remove from heat and allow to cool slightly. Halve passionfruit and scoop out pulp. Stir passionfruit pulp through warm fruit mixture and serve over sliced or halved figs, garnishing each portion with a dollop of yoghurt.

And Judah and Israel dwelt safely, every man under his vine and under his fig tree.

Old Testament: 1 Kings, iv, 25

FIGGY PEARS

4 medium pears
10 dried figs, finely chopped
½ cup carob buds (from health food shop) or chocolate buttons
juice of ½ lemon
2 tspns mixed orange zest and pulp
4-6 tbspns apple juice
2 tbspns brandy (preferably apple, or Calvados)
vanilla sugar and cinnamon

For toothache: Rub the fat of figs on the teeth, and crush small red roses and lay them over it.

from *Medicina di Quadripedibelus*
(The Book of Cures)

Set oven at 180°C/350°F. Core the pears. Combine chopped figs and carob or chocolate and fill core cavities of pears. Divide lemon juice, orange zest and pulp between pears and then place them in a small baking dish. Mix together brandy and apple juice and pour over pears. Sprinkle liberally with vanilla sugar and cinnamon;

Cleopatra's favourite fruit was the fig, and the asp with which she killed herself was brought to her in a fig basket.

According to ancient tradition, the fig leaf was a symbol of justice.

According to Greek tradition, between Advent and Epiphany goblins called *Kallikantzaroi* with squinty red eyes and hairy little bodies emerge from tunnels beneath the earth to pester human folk. However, they may be cajoled into good behaviour with gifts of dried figs and honey cakes for, like the Greek people themselves, these goblins have a sweet tooth.

cover and cook in oven for 30-35 minutes or until pears are tender. Serve hot or cold with ice cream, custard or cream.

MOROCCAN FIG SALAD

50g/2 oz dried peaches
225g/8 oz dried apricots
10g/4 oz dried prunes
50g/2 oz dried figs
50g/2 oz split almonds
50g/2 oz pine nuts
50g/2 oz sunflower seeds
2 tbspns rose water
seeds from 3 cardamom pods

Wash all the fruits and place them in a large bowl. Mix in the nuts and seeds and just cover with water. Sprinkle over the rose water and cardamom seeds. Let the fruits soak for at least 6 hours, or overnight. Much of the liquid will be absorbed, plumping up the fruits; the remainder serves as a lovely syrup.

FIG MARSHMALLOW FREEZE

16 marshmallows
2 tbspns apple juice
juice ½ lemon
6 ripe figs, peeled and chopped
3 tbspns glacé cherries, chopped
2 tbspns chopped preserved ginger
1 tbspn mixed nuts, chopped
1½ cups stiffly whipped cream

Cook marshmallows, apple and lemon juices over low heat in a non-aluminium saucepan until marshmallows are melted. Add figs, cherries, ginger and nuts, and stir well; remove from heat and allow to cool slightly. When slightly set, fold mixture into cream, then pour into metal freezer trays and freeze. Mound scoops or cubes of frozen mixture in individual serving dishes.

Fermented figs are made into an alcoholic beverage which is much prized by the Arabs. In France, a similar drink known as *Vin de figue* or *figuette* is taken as a medicine.

CHOCOLATE FIG MOUSSE

200g/7 oz dark cooking chocolate, chopped
50g/2 oz demerara sugar
2 eggs
1-2 tbspns Marsala
50g/2 oz dried figs, chopped
50g/2 oz chopped walnuts
300ml/10 fl oz cream, lightly whipped
walnut halves and thinly sliced fresh figs, to garnish

Combine chocolate, sugar, eggs and Marsala in a double boiler over low heat until chocolate is melted. Remove from heat and stir through figs and walnuts; allow to cool slightly. Whisk in the cream, mixing carefully. Spoon into serving glasses and chill 1½ hours or longer, till firm. Garnish with walnut halves and thinly sliced fresh figs, or with an extra sprinkling of grated chocolate.

FIGS FLAMBÉE

20 tiny firm green figs
6 tbspns curaçao
4 tbspn Dubonnet
200ml/7 fl oz chantilly cream (whipped cream sweetened with caster sugar and a dash of vanilla essence)

Wash figs and prick lightly all over with a fork, being careful not to split skin. Arrange them in a baking dish. Combine curaçao and Dubonnet over low heat, pour over figs and set alight. Serve immediately with chantilly cream.

FIG PUDDING WITH ORANGE SAUCE

Pudding

100g/3 oz dried figs, minced
275ml/10 fl oz orange juice
2 tbspns lemon juice
1 tspn each finely grated orange and lemon zest
1 tbspn butter
1 tbspn plain flour

Small Boy: What did you have for Dinner?
Self: Turkey, and a blazing Pudding.
Small Boy: Was it nice?
Self: It was not made on earth.
Dylan Thomas, *Conversation about Christmas*

½ tspn each cinnamon, grated nutmeg and ginger
2 eggs, separated
1 tbspn caster sugar
Sauce
1 cup water
3 tbspns sugar
1 orange
few thin slices of lemon
1 tbspn ginger in syrup, finely chopped
light sour cream, to garnish

Combine figs, orange juice, lemon juice and zests in saucepan over low heat for 10-15 minutes or until fruit softens. In a separate saucepan, melt butter, add flour and spices, and whisk together till smooth; add fig mixture, stir well and simmer for 2-3 minutes till mixture thickens. Remove from heat and cover; allow to cool slightly.
Set oven at 180°C/350°F. Lightly grease an ovenproof dish. Beat egg yolks into cooked fig mixture, then whisk egg whites with sugar till stiff and fold them gently through pudding mixture. Spoon into dish and bake for 30-40 minutes.
To make syrup, cook water and sugar gently for 4-5 minutes. Peel thin julienne strips from orange rind and add to syrup. Discard pith and seeds from orange, then cut fruit into thin slices and dice; add to syrup with lemon slices and ginger in syrup. Serve warm with slices of pudding and garnish each serving with a dollop of light sour cream.

FIG AND MANDARIN SOUFFLÉ

425g/15 oz mandarin segments in syrup
(1 large can or 2 small ones)
orange juice
dash Grand Marnier
½ tspn cinnamon
1 tspn finely grated orange zest
2 tbspns gelatine
3 tbspns caster sugar
3 eggs, separated

For diarrhoea: Pound birthwort and black dried herbs, and the fat of kome, and figs, and apply to the region of the navel, and the looseness will be stopped.
From *The Compleat Housewife* (1760) (The Book of Cures)

50g/2 oz flaked almonds, lightly toasted
100g/3 oz dried figs, minced
whipped cream, to garnish

Drain mandarin segments and in a bowl combine mandarin syrup with enough orange juice to give 2½ cups of liquid; add dash of Grand Marnier, the cinnamon and orange zest. Sprinkle gelatine over and allow to stand for 30 minutes. Transfer mandarin/orange liquid to saucepan, add 1 tbspn of the sugar and heat gently. Quickly beat egg yolks in separate bowl and, as soon as mandarin/orange liquid begins to boil, remove from heat and whisk into egg yolk mixture; leave to cool slightly.
In separate bowl, whisk egg whites with remaining 2 tbspns of sugar until stiff. Carefully fold almonds and figs through mixture then gently combine with mandarin/egg yolk mixture. Spoon complete mixture into serving dishes and place in refrigerator to set. To serve, pipe whipped cream over soufflés and decorate with reserved mandarin segments.

FIGS WITH SHERRIED MACAROON CREAM

12 ripe purple figs
4 tbspns sherry
325ml/10 oz double cream
250g/8 oz coconut macaroons
4 tbspns flaked almonds, toasted

Wash and dry figs and cut downwards into sixths, folding back segments to form a flower shape. Whip together cream and sherry and set aside.

Crush macaroons by placing in a bag and pulverising with a rolling pin (alternatively, process in food processor). Fold macaroon crumbs into whipped cream and, using piping bag with large, fluted nozzle, pipe decoratively into centre of fig 'flower'. Garnish with toasted flaked almonds.

The Wassail of Figgy Duff
Oh Christmas is icum,
Mark ye Robin's ruddy tum,
And ye winde it-te bloweth very
rough—
Rough-rough
Now do as ye be tolde
To keep us from ye colde,
And fille us up with Figgy, Figgy
Duff,
Figgy Duff.

Festive Figs

So merrily we sing
For to make ye Wlekin ring,
Though of singing we have
had-de quite enough—
Nough-nough
My maisters, if you please,
Give us a mouldy cheese
To eat with our Figgy, Figgy Duff,
Figgy Duff.

If a cheese ye have not got,
Then a groat must be our lot
As we dance around the
Mistletoe Buff—
(Bough)-Buff!
If you haven't got a groat,
You can go and cut your throat,
And ne'er have any Figgy, Figgy
Ne'er have any Figgy, Figgy
Ne'er have any Figgy, Figgy Duff,
Figgy Duff!
Adaptation of traditional English
wassailing song
(Michael Flanders)

In earlier times, certain fruits were strongly associated with the long northern winter. Apricots, figs, sultanas and currants were harvested, dried and preserved in syrup and then eaten during the Yuletide season. Being an ancient symbol of fertility and plenty, the fig, in particular, was consumed by groups of families and friends who gathered together for warmth, companionship and to share stored food and wine. In addition to the traditional plum or 'figgy' pudding still eaten at holiday feasts, many other recipes for festive cakes and sweet breads have evolved in different cultures around the world.

We Wish You a Merry Christmas
(English traditional)
We wish you a merry Christmas
We wish you a merry Christmas
We wish you a merry Christmas
And a Happy New Year.
Chorus: Good tidings we bring
To you and your kin;
We wish you a Merry Christmas
And a Happy New Year
Now bring us some figgy pudding
Now bring us some figgy pudding
Now bring us some figgy pudding
And bring some out here.
(Chorus)
For we all like figgy pudding
For we all like figgy pudding
For we all like figgy pudding
So bring some out here.
(Chorus)
And we won't go until we've got some
And we won't go until we've got some
And we won't go until we've got some
So bring some out here.
(Chorus)

Into the basin put the figs
Stirabout, stirabout, stirabout!
Next the good white flour comes
Stirabout, stirabout, stirabout!
Sugar and peel and eggs and spice,
Stirabout, stirabout, stirabout!
Mix them and fix them and cook them twice
Stirabout, stirabout, stirabout!
Traditional English carol

TRADITIONAL MINCE PIES

In the past, mince pies were more likely to be based on actual meat mince (lamb or beef) than fruit, although the same sweet spices and nuts were added to the mixture.

Fruit Mince Filling
4 apples
30g/1 oz lemon peel
30g/1 oz orange peel
110g/4 oz sultanas
60g/2 oz dried figs, minced
110g/4 oz red and green glacé cherries
100g/4 oz crystallised ginger
50g/2 oz flaked almonds
110g/4 oz beef suet (or mixture of half butter, half lard)
225g/8 oz sugar
½ tspn each cinnamon, ginger and grated nutmeg
pinch powdered cloves
grated zest and juice of 1 lemon
½ cup sherry

Prepare all fruit and nuts—peel, core and dice apples, mince peel, sultanas, figs and cherries; chop ginger and almonds finely. Chop suet finely. Combine minced fruit and nuts in mixing bowl with suet, sugar and spices, lemon juice and zest and sherry; mix well. Spoon into sterile jars and cap securely. Mincemeat should be kept for at least 4 weeks before use.

Shortcrust Pastry
250g/8 oz flour
¼ tspn baking powder
pinch salt
1 tspn sugar
2 egg yolks
1 tbspn water
squeeze lemon juice
110g/4 oz butter

Sift flour, baking powder and salt together; add sugar. Beat egg yolks with water and lemon juice. Rub butter into flour until mixture resembles fine breadcrumbs. Add water, drop by drop, mixing to a very dry dough. Turn dough onto lightly floured board and roll out thinly, to 0.3cm/⅛ inch thick. Cut rounds for tops of pies first, then larger ones for the bottoms. Line lightly-greased tartlet tins with larger round and fill with mincemeat, either store-bought or homemade from recipe above. Glaze edges with egg white, place tops on and make a hole in the centre of each. Bake for 15 minutes in a hot oven at 220°C (425°F), then reduce heat and bake further 30 minutes. Sprinkle pies with icing sugar. Serve hot with whipped cream and custard, or cold.

FIGS AND MANGOES IN SYRUP

3 large ripe mangoes, peeled, stoned and diced
½ medium honeydew melon, seeded, peeled and balled
6 figs, peeled and quartered
6 kiwi fruit, peeled and sliced
1 tbspn chopped fresh mint
750ml water
25ml/1 oz clear honey
juice and grated zest of 1 lime
grated zest of 1 lemon
juice of ½ lemon
1 tbspn kirsch
½ tspn Chinese five spice powder
½ tspn each powdered finger, cloves and coriander
1 vanilla pod, split
few sprigs mint, to garnish

Combine water, honey, lime and lemon juice and zest, Kirsch, spices and vanilla pod in saucepan and bring to the boil; cover and allow to cool before straining. Place fruit and chopped mint in a large china bowl and pour cooled syrup over; mix gently and chill well before serving. Garnish with mint sprigs.

The milky sap of the fig tree (*Ficus carica*) is a very old herbalists' remedy for warts.

FIGS IN PORT WINE JELLY WITH BRANDIED CREAM

4 fresh purple figs, peeled and very finely sliced
150g/5 oz finely chopped dried figs
100g/3 oz chopped candied orange peel
100g/3 oz currants
2 tspns lemon zest
150ml/5 oz brandy
squeeze lemon juice
300ml/9 fl oz port wine
1 tbspn sugar
50g/1½ oz powdered gelatine
red food colouring, if desired
Brandied Cream
300ml/9 fl oz cream
1 tbspn sugar
1 tspn orange zest
3 tbspns brandy

Combine dried figs, peel, currants and zest in a bowl and pour brandy and lemon juice over; cover and refrigerate overnight. Drain fruit, reserving brandy and arrange decoratively in a ring mould with sliced fresh figs. Combine brandy, port and sugar, and then stir through gelatine and food colouring; pour liquid over fruit and refrigerate till firm. To unmould, run the tip of a knife which has been dipped into very hot water around the edge of the mould and tip onto chilled plate. Serve with brandied cream, made by whipping cream with sugar and orange zest, and then folding brandy gently through mixture.

FIG CLAFOUTI

750g/1½ lbs glacé figs in syrup
250g/8 oz glacé ginger in syrup
175ml/6 fl oz milk
100ml/3 fl oz buttermilk
3 tbspns golden syrup
1 tspn minced fresh ginger

In 17th century Europe, lovers exchanged little wicker 'courting baskets' filled with figs as tokens of their affection.

12 tbspns plain white flour
custard, or double cream, to garnish

Strain figs and ginger and chop finely, retaining syrups. Combine milk, buttermilk, syrup from figs and ginger, golden syrup, eggs, vanilla, Marsala and ginger in food processor and blend till smooth. Add flour and continue to process for 1-2 minutes, or until you have a smooth batter. Cover and set aside for 15-20 minutes. Grease a 2 litre ovenproof baking dish and pour in a thin layer of batter. Place on low heat on oventop, set at medium heat, for 1-2 minutes, or until mixture barely begins to set. Scatter chopped figs and ginger over first layer of batter then pour over the remaining batter. Bake in a preheated oven at 180°C (350°F) for 40 minutes or until fluffy and golden-brown on top. Serve warm with lashings of fresh custard.

FIG AND ALMOND CRÊPES

90g/3 oz plain flour
50g/2 oz finely ground almonds
3 large eggs
1 tspn caster sugar
250ml/8 fl oz milk
90ml/3 fl oz light sour cream
pinch salt
1 tbspn light vegetable oil
few drops almond essence
butter, for cooking

Filling

15g/½ oz butter
15g/½ oz cottage cheese
juice and zest of ½ orange
300g/10 oz strawberries, hulled and chopped
200g/7 oz dried figs, chopped
pinch each powdered cloves and cinnamon
100g/3 oz flaked almonds, toasted

In 16th century Venice, almonds, pine nuts, figs, sultanas and pistachios were commonly cooked with meat and fish dishes.

Beat together flour, ground almonds, eggs, sugar, milk, cream, salt, vegetable oil and almond essence; cover and set aside in a cool place for 1 hour. Stir batter before use. Meanwhile, to prepare filling, combine butter, cottage cheese, orange juice and zest, strawberries, figs, cloves and cinnamon in saucepan over very low heat for 3-4 minutes; set aside until ready to use.

To make crêpes, heat butter in frypan and pour in 2 tbspns of batter, swirling pan so batter spreads out thinly. When crêpe sets, flip over to cook other side. When crêpe is a light golden colour, flip it onto a plate and spoon a little of the warm fig-fruit filling onto it. Scatter toasted almonds over filling, roll it up and either serve immediately with cream, or place in greased, ovenproof dish in slow oven to keep warm as you quickly make the others.

FIG AND PECAN CRUNCH

3 large cooking apples, peeled, cored and diced
2-3 tbspns maple syrup
2-3 tbspns water
2 tbspns skimmed milk powder
280ml/9 fl oz natural yoghurt
90g/3 oz pecan nuts, crushed and toasted
90g/3 oz flaked almonds, toasted
3 tbspns chopped dried figs
4 egg whites
90g/3 oz caster sugar
3 tbspns chocolate chips

Combine apples, maple syrup and water in saucepan over low heat, cooking gently till apple is mushy and liquid is reduced; set aside to cool to room temperature.

Stir milk powder into yoghurt and then fold into the cooked apple. Mix toasted pecans, almonds and chopped figs together in separate bowl and set aside. Whisk egg whites with caster sugar till stiff. Fold nuts and figs, chocolate chips and yoghurt/apple mixture into the egg whites. Spoon mixture into plastic or metal freezer container and freeze till

In Greece, figs are picked when fully ripe and dried in the hot sun for three to four days. They are then put into a slow oven to dry out completely. To store, the figs are strung on plaits of dried grass called *vroulo* and hung in the kitchen or stone larder.

quite firm, whisking once with a metal fork as ice crystals begin to form. Take out 25 minutes before serving so mixture is firm, but not too hard.

FIG AND NECTARINE ICE CREAM

425ml/14 fl oz low fat curd cheese
285ml/10 fl oz sour cream
6 tbspns apple juice concentrate
3 tbspns clear honey
150g/5 oz dried figs, minced
50g/1½ oz hazelnuts, chopped and toasted
seeds from 2 cardamom pods, crushed
few drops vanilla essence
400g/14 oz ripe nectarines, stoned and chopped
sliced fresh figs, marinated for an hour in a little apple juice concentrate and freshly grated ginger, to garnish

Combine curd cheese, sour cream, apple juice concentrate and honey in food processor; blend until smooth. Stir in figs, nuts, cardamom and vanilla essence; pour into plastic container. Either place nectarines in food processor and blend roughly, or mash with fork. Stir fruit into cheese mixture and freeze. Take out 25 minutes before serving so ice cream is firm but not too hard. Serve with sliced fresh figs, marinated as above.

SPICED FIGS WITH RHUBARB

500g/1 lb dried figs
water
1 tspn each ginger, cinnamon and nutmeg
1 bunch rhubarb, sliced into 2 inch lengths
1 green apple, peeled, cored and diced
2 tbspns maple syrup
300ml/9½ fl oz orange juice
extra 2 tspns ginger
1 orange
natural yoghurt

All the costumes since Adam's,
right or wrong,
From Eve's fig-leaf down to the
petticoat,
Almost as scanty, or days less
remote.
Byron, *The Vision of Judgement*,
St. 66

Flour of England, fruit of Spain
Meet together in a shower of rain;
Put in a bag, tied round with string;
If you tell me this riddle, I'll give
you a ring.

Traditional

Wash figs and put into a saucepan with enough cold water to cover. Add ginger, cinnamon and nutmeg and bring mixture to the boil; reduce heat and simmer for 15 minutes, or until figs have plumped up well. Drain figs and place in a serving dish.

Combine rhubarb, apple, maple syrup, orange juice, and extra ginger in saucepan and bring to the boil; reduce heat and simmer for 15 minutes, or until rhubarb is well softened. Drain, reserving liquid, and pour rhubarb mixture over figs. Peel thin julienne strips from orange. Discard pith and slice fruit very thinly, or dice. Add julienne strips and fruit to reserved syrup and pour over fig and rhubarb mixture. Chill before serving with natural yoghurt and a light dusting of extra cinnamon or ginger.

FROSTED FIGS

dried figs
hazelnuts or pecans
chopped, crystallised ginger
icing sugar

Slit side of soft, whole, dried figs and stuff centre with chopped nuts and a few pieces of ginger; press closed. Dip bottom of figs in sifted icing sugar and serve, bottom-side up.

CHRISTMAS PUDDING

The traditional English steamed plum pudding has hardly changed for centuries. It should be made a month or two before it is needed and, according to superstition, stirred once by each member of the family from 'east to west', in honour of the Three Kings. Each person then has three wishes, one of which will be granted. A Christmas pudding is not complete without a few lucky charms or silver coins hidden in it. More recently, we have used threepenny and sixpenny bits. (These pre-metric coins can still be bought from the Mint—beware 5c or 10c coins as they contain an alloy which might not taste as good as pure silver.)

500g/1 lb sultanas
500g/1 lb seedless raisins
175g/6 oz currants
110g/4 oz dried figs, minced
110g/4 oz candied orange peel, minced
110g/4 oz candied lemon peel, minced
110g/4 oz shelled almonds, chopped
300ml/½ pint brandy or rum
500g/1 lb butter
110g/4 oz brown sugar
8 eggs
500g/1 lb breadcrumbs
110g/4 oz flour
2 tbspns golden syrup or treacle
1 tspn each cinnamon, grated nutmeg and ground mace
1 tspn mixed spice

Prepare fruit well, picking over for stems, pips and discoloured parts. Pour brandy or rum over fruit and allow to stand, covered in a china pudding bowl for at least 24 hours.

Cream butter and sugar together then add eggs, one at a time, beating each one in thoroughly. Mix in breadcrumbs, flour, syrup and spices, continuing to mix well. Tie securely in well-floured pudding cloth, allowing room for pudding to swell (the cloth should be tied loosely, not snugly around the pudding). Suspend the pudding in a large pan of water so that it is submerged but not touching the bottom. For example, tie the pudding to a large wooden spoon resting across the pan. Keep boiling constantly for 6-8 hours. (Alternatively, the mixture may be spooned into a mould, covered tightly with absorbent paper and secured with rubber bands, placed in a baking dish containing boiling water, and steamed in a 180°C/350°F oven for 6 hours or so.)

VISIONS OF SUGAR PLUMS...

These are charming, old-fashioned baubles for the tree, as well as being delicious sweetmeats.

Dame, get up and bake your pies,
Bake your pies, bake your pics,
Dame, get up and bake your pies,
On Christmas Day in the morning.
Traditional nursery rhyme

500g/1 lb dried figs
225g/7½ oz dark semisweet chocolate, coarsely grated
225g/7½ oz almonds, ground and toasted
500g/1 lb icing sugar
tulle, cut into squares
fine ribbon or tinsel

Steam figs for 50 minutes to 1 hour, or until they have softened and plumped up slightly. Drain while they are still warm, cut a slit in the base of each fig and stuff with a half-and-half mixture of the grated chocolate and the almonds. Scatter icing sugar thickly on a plate and roll each stuffed fig two or three times. Cover with cling wrap, set aside and allow to cool thoroughly. Wrap each 'sugar plum' in a tulle square, tie with ribbon or tinsel and hang from the tree.

TWELFTH NIGHT CAKE

Tradition states that all decorations, including the tree, should be taken down and burnt on Twelfth Night, to avoid bad luck. In olden days, a refreshment on this occasion was the heavily-spiced Twelfth Night cake.

225g/8 oz butter
200g/7 oz white sugar
1 tbspn brown sugar
1 tspn ground ginger
1 tspn ground cardamom
1 tspn ground coriander
½ tspn grated nutmeg
4 eggs, beaten
2 tbspns sweet sherry
225g/8 oz sifted white flour
½ tspn baking powder
225g/8 oz currants
30g/1 oz chopped candied peel
30g/1 oz minced dried figs
60g/2 oz chopped walnuts

On his travels to the south of France, American president Thomas Jefferson was said to have been most enamoured of the raisins and figs from Smyrna that he ate at Marseilles. Before his return to the States, he ordered his French steward to pack a good supply of them, along with locally produced dried peaches and plums.

60g/2 oz sultanas
1 tspn each candied orange and lemon peel

Preheat oven to 120°C/250°F. Butter a large square cake tin and line with buttered, greaseproof paper. Cream butter and both sugars together. Beat in spices, eggs and sherry. Gradually mix in flour and baking powder, mixing well. Add currants, candied peel, figs, walnuts, sultanas and orange and lemon peel. Turn mixture into tin and distribute evenly. Cook until top is golden and a skewer comes away clean—approximately 3½ hours. Ice the cake with marzipan.

DUTCH CHRISTMAS BREAD

1 cup milk, warmed
¾ cup sugar
1 cup butter, softened
1 tspn salt
2 tbspns yeast
½ cup warm water
8 cups flour, sifted
5 eggs, well beaten
¾ cup sultanas
¼ cup dried figs, chopped
½ cup chopped mixed candied peel
1 cup chopped blanched almonds
½ cup chopped red and green glacé cherries
1½ tspn grated lemon zest

Grease and flour two 20cm/8 in loaf tins. Combine milk, sugar, butter and salt. In a separate bowl, combine yeast and warm water and stir until dissolved and frothy. Combine milk and yeast mixtures. Add half of the flour and the eggs and beat well. Mix in sultanas, figs, peel, almonds, cherries and lemon zest. Add remaining flour to make a soft, pliable dough. Turn dough into tins, cover with a clean towel and allow to rise until approximately doubled in size. Bake in a moderate oven 180°C/350°F for about 40 minutes, or until nicely browned. Cool on rack.

In Central Europe, roasted figs are used to flavour coffee in much the same way as chicory is used in France.

OTHER TRADITIONAL FIG RECIPES

FIG FRUMENTY

Frumenty, or fruit cereal with plenty of sugar and spices, was a popular dish in Tudor times. London vendors sold it hot, by the plateful, to passersby and it was also a traditional recipe to prepare for strangers who might call on Christmas Eve. In some parts of England frumenty was thought to be favoured by the 'farisees' or fairies, and bowlfuls were left by the back door to prevent fairy mischief during the night. Hulled or coarsely ground oats or barley or nutty-flavoured brown rice may all be substituted for the wheat—in much the same way as we blend our own muesli these days, there are no absolute rules governing the preparation of frumenty.

kibbled wheat
water
chopped candied peel
chopped dried figs, apricots and currants
milk
grated nutmeg, allspice and cinnamon, to taste
brown sugar
knob of butter

Simmer the wheat in water until it is well softened; strain, then return to the saucepan with approximately a third the amount of chopped dried fruit and peel. Add enough milk to form a porridge-like consistency and cook for a further 30 minutes, adding spices and sugar to taste. To serve, stir a knob of butter through mixture and garnish with cream. To be really authentic, add a good splash of rum to the cream.

ALBANIAN FIG COMPOTE

Known as *Turli fruta* or, literally, 'all fruits', this is a delicious summertime dessert with its dainty pink-tinged dressing.

According to Indian folklore, the wild fig only flowers for a few minutes on one night of the year. If you are patient and lucky enough to witness this, and to cut off one of the flowers as it opens, it will act as a powerful charm for you, helping to determine the whereabouts of spies and ward off evil from your family and friends.

1 cup each strawberries (hulled and sliced), raspberries, cherries (pitted and halved), blueberries, figs (peeled and sliced)
2 tbspns chopped almonds
1 tbspn sultanas
150ml/5 fl oz milk
2 cups natural yoghurt
3 tbspns sugar
1 tbspn honey

Pick over fruit, then place in serving bowl(s), with almonds and sultanas. In separate bowl mix together milk, yoghurt, sugar and honey and beat well with a whisk. Pour over fruit and refrigerate for 2-3 hours before serving.

RUMANIAN WEDDING SOUP

With figs being popularly acknowledged as a symbol of fertility and plenty, it is unsurprising they feature in this rich broth which is traditionally served at weddings. It is known as *Reiskachen Mit Huhn*.

4 carrots
1 stick celery
1 bunch shallots
4 leeks
3 tbspns chopped parsley
2 tbspns chopped tarragon
lightly salted water, to cover
1.5kg/3 lb chicken pieces
1 cup dried figs, chopped
1 cup raisins
½ cup sugar
1 cup white wine
1 lemon, thinly sliced
150g rice
3 tspns butter
2 tspns flour
1 egg, beaten
½ cup sour cream

For backache: take raysings (raisins) of the sunne 8 ounces, figgs 8 ounces, of mustard seedes 4 ounces, grind with strong wine or vinaiger till it be thicke as a plaster and spread it on leather and lay it to the grief and it will ease the pain well approved.

from *Medicina di Quadripedibus*

Clean and prepare all vegetables—peel and chop carrots, slice celery, peel and chop shallots and leeks. Place in saucepan with parsley and tarragon and enough lightly salted water to barely cover; bring to the boil, then reduce heat, cover and cook for 30 minutes over low heat. Add chicken pieces and cook for a further 45-50 minutes.

Meanwhile, combine figs, raisins, sugar and wine and cook until fruit plumps up well.

When chicken is done, add lemon slices and cook for a further 2 minutes. Remove chicken and set aside; strain stock and discard vegetables. Return stock to pan and add rice; bring to the boil and cook for 8-10 minutes.

In separate saucepan, melt butter and stir in flour till smooth and pasty; add figs in wine and continue to stir. Combine fig mixture with chicken and rice soup, and then blend in beaten egg and sour cream. Serve immediately, either as a soup or as a sauce poured over reserved chicken pieces.

BULGARIAN BAKED FIGS WITH CHERRIES AND CINNAMON

1kg/2 lb figs
4 tbspns honey
4 tbspns canned black cherries, drained and halved
3 tspns powdered cinnamon
½ cup ground almonds
½ cup reserved cherry syrup
butter

Cut off 'lids' from stem end of figs; scoop out and reserve flesh. Chop flesh and mix with honey, cherries, cinnamon, almonds, and cherry syrup. Fill figs with mixture, leaving a small hollow in the top of each; into this, place a tiny pat of butter. Replace lids, put figs in lightly buttered casserole or baking dish and pour any additional cherry syrup over them. Cook at 225°C (440°F) for 30 minutes. Serve either hot or cold with natural yoghurt or a half-and-half cream and sour cream combination, dusted with cinnamon.

Figs were considered a symbol of fertility, luck and prosperity by the ancient Greeks. They also named figs a 'pure food' suitable for pregnant women, along with cheese and barley cakes.

GREEK FIG AND ORANGE SALAD

This is a variation on a dish known as *Portokalia me krema milo*, which is served in Salonika.

6 ripe oranges
2 crisp apples
4 purple figs, peeled and chopped
½ cup rum
2 tbspns sugar, or to taste
4 tbspns finely chopped almonds
1 tspn grated orange zest
150ml/5 fl oz cream

Cut 'lids' off stem ends of oranges and hollow out flesh; discard pith and pips and chop flesh. Peel, core and grate apples and mix with orange pieces, fig flesh, rum, sugar, chopped almonds and orange zest. Whip cream and fold through fruit mixture before piling into hollowed-out oranges. Refrigerate before serving.

GREEK FIGS IN WINE

This simple, delicious dish is known as *sika* and is traditionally served on a bed of fig leaves and accompanied by strong, piping hot black coffee.

In rural areas in Greece, figs are dried and preserved in many different ways. One method is to slice through to the bottom of the fig, pressing back the four segments and spreading them flat to dry in the sun. When it is nearly dry, but still slightly sticky on the flesh side, it is stuffed with chopped nuts and a second fig is pressed flat on top before the whole sweet is oven-dried.

fresh green figs
white wine
caster sugar
fig leaves, if available, to garnish

Wash figs, pat dry and, using a fork, lightly prick each fruit a few times. Place figs in large non-aluminium container—preferably earthenware—with enough wine to barely cover; set aside in a cool dark place overnight. Strain figs, add sugar to wine to taste and cook in a non-aluminium saucepan to form a syrup; pour over figs and refrigerate till well chilled. Serve on a plate garnished with fig leaves.

To Make Syrop of Hysop for Colds: Take an handful of Hysop, of Figs, Raisins, Dates, of each an Ounce, French Barley one ounce, boyl it therein three pintes of fair water to a quart, strain it and clarifie it with two whites of Eggs, then put in two pound of fine sugar and boyl it to a Syrop.

The Queen's Closet Opened, by W.M., Cook to Queen Henrietta Maria (1655)

RUSSIAN EASTER CHEESE

Known as *Paskha*, this is a variation of the special cheese which is served to honour patron saints' days in Russia. It was once traditional to make the Easter cheese in a special china or earthenware dish which was decorated inside with raised crosses. Thus, when the cheese was turned out, there were crosses imprinted on both sides. To be authentic, the cheese should also be decorated with a ring of small paper rosettes.

400g/13 oz cottage cheese
125g/4 oz cream cheese
2 tbspns sour cream
2 tbspns butter
1 tspn vanilla essence
1 cup ground almonds
6 egg yolks
1 cup caster sugar
zest of 2 lemons
3 tbspns finely chopped orange zest
½ cup chopped dried figs
½ cup sultanas
½ cup chopped crystallised ginger and glacé cherries
For decoration: slivered almonds, fresh figs and glacé cherries.

Place cottage cheese in muslin cloth and hang up to drain overnight. Process cottage cheese, cream cheese, sour cream, butter, vanilla, ground almonds, egg yolks and sugar, lemon and orange zest in a blender on low speed until smooth. Mix in figs, sultanas and fruit with a spoon and put mixture in a glass dish. Decorate with almonds, sliced figs and whole glacé cherries.

In medieval times, bay leaves were used in sweet dishes. *The Goodman of Paris* (1393) describes a recipe of '...cooked apples and large Provencal figs roasted, with bay leaves thereon.' Today, dessert figs are still often packed with a bay leaf to give a slight flavour and aroma.

BULGARIAN LAMB STEW

250g/8 oz finely chopped dried figs
250g/8 oz finely chopped dried apricots
4 tbspns butter
2 quinces, peeled and sliced
1 medium sized onion, peeled and minced
½ tspn powdered cloves
500g/1 lb baby lamb, boned and cubed

Ancient fertility and harvest customs, such as those of the Druids, often featured the ceremonial preparation and eating of figs. At these feasts, blessed water was borne in pewter vessels crowned with fig leaves, so potent was this fruit considered in giving health, life and fecundity.

Place figs and apricots in a bowl with enough water to barely cover and set aside to soak for two hours; drain and reserve liquid. Melt butter in heavy-based saucepan and cook quinces for 2-3 minutes; add onion, cloves and meat and continue to cook over medium heat for 8-10 minutes, or until meat is well browned. Reduce heat, add figs and apricots and continue to cook for 45 minutes-1 hour, adding a little of the reserved water from soaking the fruit, if necessary.

RUMANIAN FIG DUMPLINGS

4 cups plain flour
½ tspn salt
3 eggs
2 tbspns water
1 cup chopped glacé figs
1 tbspn chopped, candied lemon peel
1 tbspn fig jam, or marmalade
1 tbspn brandy
icing sugar, for sprinkling

Sift together flour and salt; add eggs and water and mix to form a soft dough. Roll out dough and cut rounds about 3 inches in diameter. Mix glacé figs, peel, jam or marmalade and brandy together in a bowl and place teaspoonfuls in the centre of each round; pull up edges of each and press together with moistened fingertips. To be authentic, dumplings should be cooked in sweetened, boiling water for 10-15 minutes; alternatively, fry in hot oil or fat and sprinkle with icing sugar to serve.

BULGARIAN FIGS WITH CHICKEN AND CHESTNUT STUFFING

12-16 large figs
oil or butter, for frying
1 onion, peeled and minced
1 tbspn flour
1 tspn paprika
½ tspn cinnamon
3 cups light chicken stock
4 chicken breasts, diced

Full on its crown,
A fig's green branches rise,
And shoot a leafy forest to the skies.
Homer, *Odyssey*. Bk xii, 1 125

The ancient Egyptians believed that the goddess Netpe gave 'a fig of the sycamore' to the newly dead to carry as they entered 'eternal bliss'.

500g/1 lb chestnuts
1 tbspn sugar
1 tbspn tomato paste
sherry or water, for cooking

Cut off 'lids' from stem ends of figs; hollow out fruits until only a ¼" thick shell is left. Reserve flesh and chop; set aside. Heat butter or oil and fry onion till transparent; stir in flour, paprika and cinnamon, then add one cup of stock. Add diced chicken, cover and cook over low heat for 25-30 minutes. Meanwhile, cut open chestnuts and boil in lightly salted water for about 15-20 minutes; drain, peel and remove inner skin; chop chestnut meat. Add chestnuts, fig flesh, sugar and tomato paste to chicken mixture. Fill fig shells with this mixture, replace lid and place stuffed figs side by side in a wide flat pan. Pour in rest of chicken stock, adding sherry or water to cover figs ¼ the way up. Bake in a slow oven for 35 minutes. Serve with wild rice and steamed zucchini or carrot strips.

FIG-NUT BONBONS

These sweetmeats are traditionally served to guests in Balkan countries, such as Rumania and Bulgaria.

2kg/4 lb ripe purple figs
1kg/2 lb walnuts, shelled and skinned

Wash figs and cut a slit to the centre of each; tuck a whole walnut in the centre of each fig and press slit closed. Arrange figs stem end down in a large, flat, baking sheet and place in a very slow 120°C (240°F) oven for 1½-2 hours for three days running or until quite dried. Store in an airtight container.

According to ancient tradition, the fig leaf was a symbol of justice; so, in the days before everyone could read, a lawyer would hang a carved or painted fig leaf outside his offices to indicate his whereabouts.

TRADITIONAL LENT PIE

In times gone by, the fasting period of Lent was strictly observed and meat dishes were temporarily replaced by spicy fruit ones, such as this one.

Along with the grape vine, the fig has been used as an emblem of Israel.

6 apples, peeled, cored and chopped
2 quinces, peeled, cored and chopped
1 cup stoned dates, chopped
1 cup dried figs, chopped
1 tbspn currants
2 tbspns mixed candied peel
½ tspn each powdered ginger, cloves, mace and nutmeg
1 tbspn rosewater
juice of 1 orange
1 tspn grated lemon zest
2 tbspns sherry
2 tbspns brandy
450g/1 lb puff pastry
melted butter

Combine apples, quinces, dates, figs, currants, peel, spices, rosewater, orange juice, lemon zest, sherry and brandy in a large mixing bowl. Preheat oven to 220°C/420°F. Roll out half the pastry into a circle and use to line pie dish. Prick pastry shell and brush with melted butter before piling fruit mixture into the centre. Roll out remaining pastry and cut to form a lid for the pie, pinching edges together and trimming with a knife. Use fork to pierce decorative pattern in pie crust so steam can escape. Brush top of pie with remaining butter and bake in oven till pastry is golden (40-50 minutes).

HUNGARIAN FIG WINE

Known as *filligebor*, this is a traditional recipe which produces a sweet, strong wine.

1kg/2 lb ripe purple or green figs, halved
50g/1½ oz yeast
¼ cup cream of tartar
2kg/4 lb white sugar and 1 kg/2 lb brown sugar, mixed together
3 cups water

Place figs in large ceramic or earthenware preserving jar with half the yeast and ⅓ cream of tartar. Boil 1kg/2 lb sugar mixture with 1 cup of water to produce a syrup; pour this over figs, cover and set aside in a cool, dark place for two weeks.

Drain figs and reserve liquid. Repeat process twice more, reserving liquid again both times. Combine the three batches of liquor and bottle, corking securely.

On the Greek island of Cythera, the first green figs of the season are picked when they are no bigger than the first joint of a man's thumb. These tiny figs are preserved in syrup or dried and are usually served stuffed with a blanched almond.

A Celebration of Cakes and Homemade Sweets

Figgy passions

The proper way to eat a fig in society is to split it in four, holding it by the stump, and open it so that it is a glittering rosy moist honeyed heavy-petalled, four-petalled flower. Then you throw away the skin after you have taken off the blossom with your lips. But the vulgar way is just to put your mouth to the crack and take out the flesh in one bite.

The fig is a very secretive fruit. The Italians vulgarly say it stands for the female part, the fig fruit; the fissure, the yoni. The wonderful moist conductivity towards the centre, involved, inturned, one small way of access only and this close curtained from the light. Sap that smells strange on your fingers, so that even goats won't taste it. And when the fig has kept her secret long enough, it explodes and you see through the fissure, the scarlet. And the fig is finished.

Nothing tastes as good as the cakes, biscuits and cookies you bake yourself—and there is the added pleasure of filling your kitchen with the delicious aromas of spices like ginger and cinnamon. Enjoy these delectable treats either as snacks, or to accompany coffee after a meal.

FIG FANCIES

1 cup dried apricots
1 cup dried figs
¼ cup prunes, stoned
¼ cup sun-dried pineapple
1 cup maple syrup
1 tspn rum
dash lemon juice
½-¾ cup ground almonds
½ cup toasted shredded coconut and ¼ cup carob powder, mixed together on plate, for rolling

Place apricots, figs, prunes and pineapple in food processor and chop roughly. Combine in mixing bowl with maple syrup,

The year is over. That's how the fig dies, showing her crimson through the purple slit. Like a wound, the exposure of her secret in the open day; like a prostitute, the bursting fig makes a show of her secret.

D. H. Lawrence, *Women in Love*

rum, lemon juice and enough ground almonds to form a workable texture. Roll spoonfuls into balls in the palms of your hands, then toss well in coconut and carob powder mixture. Chill fancies thoroughly before serving; may be stored between layers of greaseproof paper and frozen.

RICH FRUIT CAKE

500g/1 lb butter
500g/1 lb sugar
9 large eggs
3 additional egg yolks
500g/1 lb flour
pinch salt
500g/1 lb raisins
250g/8 oz currants
225g/7½ oz dried figs, minced
110g/4 oz dried apricots,chopped
110g/4 oz citrus peel
110g/4 oz lemon peel
225g/8 oz candied cherries
110g/4 oz coarsely chopped almonds
lemon essence

Preheat oven to 200°C/400°F. Butter two 20 cm/8 in cake tins and line with greaseproof paper; butter again and dust very lightly with flour. Cream butter and sugar. Beat eggs and add slowly to butter mixture. Add sifted flour and salt, then fruit, peel, nuts and lemon essence. Place batter in tins and put in baking dish with about 4cm/1½ in hot water; bake for 3½-4 hours, or until cakes spring back to touch. Cool cakes in pan, then cool completely on rack.

Famous figs

Ancient history tells us the fig may have been the fruit which caused the Romans to destroy the city of Carthage. In a debate in the Roman Senate, Cato was said to have picked up a fresh ripe fig, in prime condition. 'When do you think this fig was picked?' he thundered. 'Only three days ago, this fig was hanging from the tree and it comes from Carthage. Judge from this how close the enemy is!'

CRISPY FIG SHORTBREAD

150g/5 oz butter
75g/2½ oz caster sugar
grated zest of 1 orange
2 eggs
1 tbspn sherry
1 tspn cinnamon sugar
50g/1½ oz ground hazelnuts

In Greece, the *zaharoplastis*, or sweet vendor, is a common sight. His tempting displays will nearly always include *halvas*, honey-cakes and glacé fruits, including small, whole figs which have been preserved carefully so that neither their shape nor colour is lost.

350g/12 oz plain flour, sifted
pinch each salt, nutmeg and cinnamon
250g/8 oz fig jam
whipped cream, quartered fresh figs, mint sprigs and orange curls, to garnish

Cream butter and sugar together; add orange zest, eggs, sherry, cinnamon sugar and hazelnuts. Stir in flour, salt, nutmeg and cinnamon, mixing well to form a smooth dough. Divide dough into two pieces, wrap in plastic cling film and freeze for one hour. Set oven at 180°/350°F. Remove one piece of dough from freezer and grate coarsely; spread grated dough thickly and evenly over the base of a greased baking dish. Spoon fig jam evenly over grated dough, then grate other piece of dough and spread that over the top. Bake for 35-40 minutes, or until top is golden and crispy. Serve with lashings of whipped cream and garnish with sliced fresh figs, mint sprigs and orange curls.

FIG AND WALNUT GINGERBREAD

250g/8 oz plain flour
1 tspn bicarbonate of soda
1 tspn cinnamon
½ tspn nutmeg
1 tbspn powdered ginger
50g/1½ oz dried figs
50g/1½ oz crystallised ginger, chopped
50g/1½ oz walnut halves, chopped
finely grated zest of 1 lemon
100g/3 oz butter
75g/1½ oz honey
150ml/3½ fl oz milk
2 eggs, beaten
extra 50g/1½ oz walnut halves, for garnish

A superstition from Brittany in France to determine how long a person will live: take a fig leaf and, pressing very gently with an embroidery needle, write that person's name on the leaf's surface. If the leaf lasts a week, the person in question will live a long life; if it withers and dies overnight, it is not a good omen.

Soak figs in hot water for 45-50 minutes till they are plumped, then drain and chop finely. Lightly butter a slice pan and set aside. Preheat oven to 180°C (350°F).
Sift flour, bicarbonate of soda and spices together and mix with figs, ginger, chopped walnuts and lemon zest. In a small saucepan, mix together butter, honey and milk over low heat.

Indian Vegetable Syrup for Coughs:
Boil the juice of sweet pomegranates and figs together, and let the patient drink the mixture for three days. Or, take three eggs, and goat oil and honey, and a few husked peppercorns, mix them together, make hot, and use as a liniment.
from *The Compleat Housewife*, 1760

Allow to cool slightly then add to fig mixture, beating well. Add eggs, continuing to beat vigorously until mixture is very well blended and quite smooth. Pour into pan and top with walnut halves. Bake in oven for 45 minutes or until a skewer inserted into centre of gingerbread comes away clean.

CARIBBEAN CAKE

450g/1 lb butter
450g/1 lb sugar
4 eggs
650g/23 oz flour, sifted
2 tspns bicarbonate of soda
1 tspn powdered ginger
½ tspn grated nutmeg
2½ cups milk
450g/1 lb currants
225g/7 oz raisins
110g/3½ oz slivered almonds
110g/3½ oz chopped dried figs
2 tbspns rum

Set oven at 180°C/350°F. Cream together butter, sugar and eggs; stir in half the flour and the bicarbonate of soda, ginger and nutmeg, then the milk, mixing well before adding the rest of the flour. Stir currants, raisins, almonds, figs and rum through batter. Pour batter into greased cake tin and bake for 1¾ hours, or until a skewer inserted in the centre of the cake comes away clean.

FIG AND BUTTERMILK CAKE

110g/3½ oz butter
½ cup sugar
1 cup buttermilk
2 tbspns honey
2 cups flour, sifted
pinch salt
1 tspn bicarbonate of soda
½ tspn each allspice, cinnamon and grated nutmeg
1 cup finely diced dried figs
½ cup walnuts, chopped
1 tbspn chopped candied lemon peel

The fig is the Islamic 'Tree of Heaven'.

In Bari in Italy, fennel seeds are used to flavour packets of dried figs.

Risschewes de frute.
Take ffigges, and grinde hem in a morter al smal with a littel oyle, and grynde with hem, clowes, and maces; and then take hem vppe in-to a dissh, and caste thereto pynes, saundres, reisons of coraunce, myced dates, pouder of Peper, Canell, Saffron, and salt; And then make fyne paast of floure, water, sugur, saffron and salt, And make ther-of faire kakes; and then rolle the stuff in thi honde, and couche hit in ye kakes; kutte them, and so folde hem (togedrys) as risshewes (rissoles), And fry hem in goode Oyle, And serue hem forthe hote.
from *Two Fifteenth Century Cookery Books* (a collection of recipes from 15th century manuscripts, published by The Early English Text Society)

During the ancient Greek festival of *Thargelia*, a scapegoat—usually a criminal or unpopular person—was ceremonially expelled from the city and supposedly thus took all the bad luck with him. He wore a necklace of black figs, representing the men, and one of white figs, representing women. He was then beaten with wild fig boughs and leeks, which were both regarded as purgative, while flutes were played.

Set oven to 180°C/350°F. Cream together butter and sugar, add buttermilk and honey, then stir in flour, salt, bicarbonate of soda and spices and mix well. Add figs, walnuts and candied lemon peel, mixing thoroughly. Place in well greased tin and bake for approximately one hour, or until a skewer inserted in the centre of the cake comes away clean.

CHOCOLATE-FIG BUTTONS

6 tbspns flour
2 tbspns cocoa
½ tspn powdered cloves
6 tbspns icing sugar
6 tbspns butter
3 eggs
6 tbspns finely chopped dried figs
1 tbspn grated semi-sweet dark chocolate

Sift flour, cocoa and cloves together. Add sugar, then work in butter and eggs to form a dough. Mix in figs and grated chocolate, stirring well. Place heaped teaspoonfuls on a lightly buttered baking sheet, about two inches apart, and bake at 180°C/350°F for 30 minutes, or until done.

FIG-NUT MUNCH

25g/¾ oz cashew nuts
25g/¾ oz peanuts
25g/¾ oz sunflower or shelled pumpkin seeds
125g/4 oz dried figs, chopped
125g/4 oz dried apricots, chopped
50g/1½ oz dried banana flakes
25g/¾ oz dried pawpaw, chopped

Spread cashews, peanuts and sunflower or pumpkin seeds on a baking sheet and toast under a medium-hot grill for 3-4 minutes. Combine hot nuts and seeds with figs, apricots, banana flakes and pawpaw with a wooden spoon. Take spoonfuls of the mixture and press firmly together with the hands, then store between layers of greaseproof paper in an airtight container.

Among the ancients, the fig was a phallic symbol, sacred to Dionysius (Bacchus), Jupiter, Silvanus and Priapus. Oddly, it was also a female fertility charm and baskets of figs were carried in processions honouring female goddesses or the Earth Mother.

FIG AND CHESTNUT CAKE

Cake

6 egg whites
1 cup sugar
3 tbspns cocoa
1 ¼ cups ground hazelnuts
4 tbspns fig jam

Filling

4 tbspns sugar
4 tbspns fig jam
dash vanilla
2 tbspns cocoa
3 tbspns butter
3 egg yolks
6 tbspns chestnut purée
whipped cream, to decorate

Beat egg whites, sugar and cocoa together till stiff. Mix in ground nuts and jam and place in a pair of round, buttered cake tins; bake in 180°C/350°F oven for 30 minutes, or until a skewer inserted in the centre of cakes comes away clean. Allow cakes to cool thoroughly on a rack.

To make filling, combine sugar, jam, vanilla and cocoa over medium heat. Remove from heat and cool slightly before adding butter, egg yolks and chestnut purée. Spread warm filling between cooled cakes and pipe whipped cream decoratively on the top.

FIG-NUT BARS

45g/1 ½ oz butter
½ cup caster sugar
2 tbspns whisky
1 tbspn liquid glucose
250g/8 oz dried figs, chopped
½ cup chopped hazelnuts
1 tbspn ground almonds
¼ cup chopped preserved ginger
1 tspn vanilla
caster sugar, for sprinkling

In ancient times, sages believed that sleeping in the shade of a fig tree would engender deep sleep and 'quiet the vexatious soul'.

So counsel'd he, and both together went
Into the thickest wood; there soon they chose
The fig-tree, not that kind for fruit renown'd,
But such as at this day to Indians known
In Malabar or Decan spreads her arms
Branching so broad and long, that in the ground
The bended twigs take root, and daughters grow
About the mother tree, a pillar'd shade
High overarch'd, and echoin walks between.
Milton, *Paradise Lost*. Bk ix, 1., 1099

Grease a slice pan thoroughly and set aside. Combine butter, sugar, whisky and glucose in a saucepan over low heat. Bring to the boil and cook for one minute, remove from heat and stir through figs, hazelnuts, ground almonds, ginger and vanilla, mixing thoroughly. Press into tin and refrigerate. When firm, cut into bars and sprinkle with caster sugar.

FIG AND POPPYSEED STRUDEL

½ cup buttermilk
400g/14 oz poppy seeds
½ cup sugar
1 tbsp lemon zest, grated
½ cup raisins
½ cup chopped dried figs
2 apples, peeled, cored and finely chopped
½ tspn cinnamon, pinch cloves
650g/1½ lb prepared strudel dough or filo pastry
melted butter
extra cinnamon, for dusting

Set oven at 225°C/440°F. Combine buttermilk, poppy seeds, sugar and lemon zest in heavy based saucepan over low heat, stirring to form a paste. Add raisins, figs, apples, cinnamon and cloves and remove from heat. Brush strudel dough liberally with butter and spread thickly with fig and poppyseed mixture before rolling up, jam-roll style. Brush top of strudel with extra melted butter and dust with extra cinnamon and/or icing sugar, if desired. Place in lightly buttered baking dish and bake for 30 minutes, or until pastry is crisply golden.

The betrayer of Christ, Judas Iscariot, is said to have hanged himself on a fig tree; this is one reason why some sources regard the fig as a tree of ill-omen. However, there is also a school of thought which says it was an elder tree that Judas used.

FAVOURITE FIG-ORANGE CANDIES

225g/8 oz wholemeal self-raising flour
1 tbspn sugar
pinch salt
110g/4 oz butter
1 tbspn chopped dried figs
1 tbspn chopped raisins
1 tbspn chopped crystallised ginger
1 tbspn finely chopped walnuts or pecans

In ancient Greece it was customary to place fresh fig boughs at the doors of a house before embarking on a journey. This would ensure luck on the trip and a safe and happy return for the travellers.

½ tbspn chopped candied orange peel
1 tbspn golden syrup
dash whisky
1 tspn orange juice
icing sugar

Sift flour, sugar and salt together, then rub butter into flour mixture. Add enough water to make a stiff dough, then mix in figs, raisins, ginger, nuts, peel, golden syrup, whisky and orange juice. With your hands, roll fruit and nut mixture, a quarter teaspoonful at a time, into balls. Bake in a 220°C (425°F) oven for 10-12 minutes. Allow to cool on racks, then roll in icing sugar while still warm.

FIG GEMS

400g/14 oz dessert figs, minced
250g/8 oz salted peanuts, crushed
½ cup raisins, crushed
juice of 1 lime
peanut butter
crushed peanuts, for rolling

Mix figs, peanuts and raisins thoroughly together in a chilled china bowl, adding enough lime juice and peanut butter to bind. Form spoonfuls of mixture into balls and roll in crushed peanuts, then chill in refrigerator; insert toothpicks and serve as a cocktail-hour appetiser.

"DEVIL'S DROPS"

This is the odd translation of the Hungarian *Ördögpirulák*, the name given to these delicious little sweets. Certainly they are extremely tempting...

1 cup sugar
4 tbspns grated orange zest
2 tbspns orange juice
2 tbspns rum
1-2 tbspns ground almonds
1 cup chopped toasted almonds
¾ cup chopped glacé figs
grated semi-sweet chocolate, for rolling

Train up a fig tree in the way it should go, and when you are old, sit under the shade of it.
Charles Dickens, *Dombey and Son*, Bk i, Ch 19

In 17th century Europe, lovers exchanged little wicker 'courting baskets' filled with figs as tokens of their affection.

Train up a fig tree in the way it should go, and when you are old sit under the shade of it.

Peel a fig for your friend
And a peach for your enemy
Traditional

Cook sugar, orange zest, orange juice and rum together over medium heat. Stir in ground almonds, toasted almonds and glacé figs, adjusting quantity of ground almonds and juice or rum until you have a mixture which may be rolled into small firm balls. Roll the balls in chocolate and serve with coffee and liqueurs.

SOUR FIG AND LEMON CAKE

2 lemons
1 tbspn freshly grated lemon zest
125g/4 oz butter
1 cup firmly packed brown sugar
3 large eggs
2 cups sifted flour
2 tspns bicarbonate of soda
¾ cup sour cream
¾ cup diced candied lemon peel
¾ cup figs in syrup, drained and chopped

Set oven at 180°C/350°F. Grease and flour cake tin and set aside. Finely grate the zest from the lemons. Seed lemons and mince flesh. Cream butter and sugar together, then add eggs and beat well. Sift together flour and bicarbonate of soda and add to butter mixture, alternating with sour cream. Stir in the candied lemon peel, figs, grated lemon zest and 1½ tbspns minced lemon flesh. Turn batter into tin and bake for about 30 minutes, or until cake springs back to a light touch. Remove from oven and allow to cool on rack.

...the different colours and textures were arranged in layers and dressed with oil, vinegar and sugar. This salad contained: raisins of the sun; blanched, shredded almonds; spinach; slices of orange and lemon; shredded red colewort and dried figs; black and green olives; hearts of lettuce; red sage leaves; pickled cucumbers; and capers...
from *A Prince's Sallet*, Gervase Markham (1659)

UPSIDE-DOWN FIG PUDDING

1 tbspn brown sugar
2 firm pears, peeled, cored and chopped
110g/3½ oz dried figs, chopped
50g/1½ oz dates, stoned and chopped
110g/3½ oz butter
110g/3½ oz honey
110g/3½ oz maple syrup
2 eggs
75ml/2½ fl oz milk
1 tbspn rum

In the name of the Prophet — figs!
Horace and James Smith,
Johnson's Ghost

75ml/2½ fl oz buttermilk
1 tspn bicarbonate of soda
225g/7 oz plain flour
½ tspn each powdered cinnamon, powdered ginger and grated nutmeg.

Set oven to 180°C/350°F. Grease an 8″ pudding bowl or cake tin with butter, then line with greaseproof paper and butter this, too. Sprinkle sugar evenly over base of bowl. Arrange chopped pears, figs and dates on the sugared base.

In a small saucepan over low heat, melt together butter, honey and maple syrup; remove from heat. Beat eggs, milk, rum and buttermilk together with bicarbonate of soda; mix in flour and spices and beat to form a smooth batter. Pour batter over fruit and cover bowl or tin with buttered greaseproof paper, secured with string or an elastic band. Place pudding bowl in a baking pan with about 1½″ water and bake 1½-2 hours. Serve hot with custard, natural yoghurt or cream.

FIG-APPLE MUNCHIES

450g/15 oz green cooking apples, peeled, cored and diced
1 tbspn apple cider vinegar
75g/2½ oz self-raising flour
100g/3½ oz dried figs, chopped
100g/3½ oz dates, stoned and chopped
3 tbspns currants
100g/3½ chopped peanuts
50g/1½ oz dessicated coconut
1 tbspn flaked almonds
50ml/1½ fl. oz apple juice
50ml/1½ fl oz dark grape juice
50g/1½ oz butter
butter or oil, for greasing slice tin

For snake bite
Apply lobelia or a bag of dampened salt or smear it with the fat of figs…
Dr Chase's Third, Last and Complete Receipt Book and Household Physician (1887)

Preheat oven to 180°C/350°F. Lightly grease an 8″ slice tin; set aside. Place apples in non-aluminium saucepan with 1 tbspn apple cider vinegar; cover and cook until fruit is fully softened, stirring well to form a thick purée. Remove from heat and stir in flour, then figs, dates, currants, nuts, coconut,

For earache—
Boil figs up in vinegar and old wine, and pour into the ears.
from *The Compleat Housewife, 1760*

almonds, apple juice, grape juice and butter. Mix well with wooden spoon till mixture is pasty, adjusting ingredients to make it more or less moist, as desired. Spread mixture out in slice tin, pressing down lightly and evenly. Bake in oven for 45 minutes or until top is lightly browned. Allow to cool in tin, then cut into bars and store in an airtight container.

FIG AND WALNUT CRESCENTS

225g/7 oz butter
½ cup sugar
½ tspn bicarbonate of soda
1 tspn coffee essence
½ tspn nutmeg
1 tspn mace
1 egg
1½ cups flour, sifted
225g/7 oz chopped dried figs
3 tbspns chopped walnuts
icing sugar, for dusting

Grease a baking sheet and set aside; preheat oven to 180°C/350°F. Cream butter and sugar together till fluffy. Dissolve bicarbonate of soda in 1 tbspn boiling water and add to mixture, then add coffee essence, nutmeg, mace, egg and flour, then the figs and walnuts, mixing well. Shape spoonfuls of mixture into crescents and bake for 10-12 minutes, or until golden brown.

COCONUT-FIG COOKIES

1 tspn baking powder
2 cups wholemeal plain flour
1 tbspn cocoa powder
½ cup butter, softened
¼ cup honey
2 tspns vanilla essence
¾ cup pineapple juice
½ cup dessicated coconut
⅔ cup chopped dried figs
¼ cup dried banana chips (from health food stores)
1-2 tbspns currants

for teeth which are loose—pound dryed figges, boil then in olive oil and rub the teeth with the mixture when hot. If thou haddest some of this in thy mouth it will strengthen every member therein.
Warren's Household Physician
(I Warren, Bradley, UK 1882)

Lightly grease a baking sheet and set aside. Sift baking powder, plain flour and cocoa powder together. Mix in butter, honey, vanilla essence and pineapple juice, beating on low speed to form a dough. Stir in coconut, figs, banana chips and currants and mix thoroughly. Cover dough and chill for at least one hour. Set oven to 190°C/375°F. Take spoonfuls of dough, roll lightly between palms to form ball shapes and flatten onto baking sheet. Bake for 20-25 minutes or until golden.

CRAZY FIG AND PECAN POPCORN

6 cups popped corn
⅔ cup roughly chopped dried figs
2 cups pecan halves
⅔ cup almonds
1⅓ cups sugar
1 cup butter
½ cup golden syrup
1 tspn vanilla

Combine popcorn, figs and nuts and spread out on baking sheets. Combine sugar, butter and golden syrup over medium heat; bring to the boil, stirring constantly, until mixture starts to caramelise. Stir in vanilla. Remove from heat and pour over corn and nuts, mixing around so all are thoroughly coated. Allow spread-out mixture to dry as a solid mass, then snap into pieces. Store in airtight container away from heat.

SPICED FIG CAKE

Thin slices of rich cakes like these are traditionally served in Italy with small cups of very strong espresso coffee.

200g/7 oz peeled almonds, toasted and chopped
200g/7 oz hazelnuts, toasted and chopped
200g/7 oz mixed candied fruit, chopped
200g/7 oz dried figs, chopped
¼ tspn each ground coriander, mace, powdered cloves, nutmeg and allspice
1 tspn cinnamon

In *The Mysteries of Isis,* the ancients believed the fig to be emblematic of law and justice for its leaves 'conceal...and cover the fruit as words conceal the sense of the law...'

For the Italian dish of prosciutto ham and figs to be authentic, the ham slices should be wrapped loosely around iced, peeled, fresh figs and be offered on a chilled platter.

50g/1½ oz plain flour
75g/2½ oz cocoa
150g/4½ oz golden syrup
150g/4½ oz sugar
10 sheets rice paper
icing sugar and cinnamon, to garnish

Set oven to 180°C/350°F. Combine nuts, fruit and spices in large mixing bowl. Sift together flour and cocoa and combine with fruit mixture. Melt together golden syrup and sugar in double boiler over low heat until mixture reaches 'soft ball' stage (when mixture may be rolled into soft balls between wetted fingers). Pour this syrup into fruit mixture and combine thoroughly, using either your hands or a wooden spoon for best results. Place sheets of rice paper on top of buttered baking sheet and form mixture into a large round about 1″ thick on top of them. Bake for 30-40 minutes; when cool, trim off excess rice paper and dust with icing sugar and powdered cinnamon.

The Four Mendicants
Les Quatre Mendiants
The first four orders of mendicant friars established themselves in 12th century France during the reign of Louis the Fat. They were the Dominicans or Black Friars, the Carmelites or White Friars, the Franciscans or Grey Friars and the Austin or Augustinian Friars. These gentlemen became famous for their begging techniques, the way they sucked the labouring classes dry. Anything, however poor, was good enough for their wallets.

When it came to Lent, these revered holy men kept such foods as black raisins, white almonds, grey figs and hazelnuts for spinning out their meals, and helping them to get over the taste of salmon, trout and so on, which, to mortify their flesh, they were obliged to eat on fasting days.

It's for this reason, so they say, that a wit seeing these dried fruits and nuts together on a plate one day, cried out: 'Look' the four begging friars—les quatre mendiants!'

from *Historiographie de la Table*, by C Verdure, published in 1833 quoted in *Jane Grigson's Fruit Book* (Penguin, Great Britain, 1982)

FIG AND ROSEWATER CAKE
175g/6 oz dried figs
125g/4 oz butter
½ cup caster sugar
3 eggs
grated zest and juice of 1 orange
210g/7½ oz self-raising flour
125g/4 oz semolina
¼ cup rosewater
Syrup
¼ cup caster sugar
½ cup hot water
1 tspn rosewater
icing sugar, to garnish

Cover figs with boiling water and stand for 35-40 minutes or until well softened; drain thoroughly and chop finely. Set oven to 180°/350°F. Beat butter and sugar together till creamy then add eggs, one at a time, beating each one in well at moderate speed. Beat in orange juice and zest, then

Sicilian orchardists would once bind the trunks of their fig trees with young elm branches. This, they believed, prevented the fruit from falling or being damaged in any way before it was perfectly ripe.

flour, semolina and rosewater. Stir in figs and pour batter into greased cake tin. Bake for 45 minutes, or until a skewer inserted in cake comes away clean. Allow cake to cool on rack to room temperature, then pour over syrup, made by combining sugar, water and rosewater over moderate heat till sugar is quite dissolved. Dust with sifted icing sugar.

ALGERIAN FRUIT AND NUT SLICE

This is a variation of the dish known as *Halva ditzmar*; it is very easy to make and it keeps well, too.

440g/15½ oz dates, stoned and chopped
440g/15½ oz figs, peeled and pounded
225g/7 oz blanched almonds, lightly toasted and then ground
2 tbspns orange juice
100g/2½ oz candied orange peel, chopped
zest of 1 lemon, finely chopped
1 tspn aniseed
1 tbspn white sugar
½ tspn cinnamon
orange blossom water, if available
2 tbspns honey

Combine all ingredients in bowl, mixing well. Press into a lightly oiled metal cake tin and refrigerate to firm. Store in an airtight container, cutting off slices as required.

The colony is justly famed for the goodness and variety of its fruits: Peaches, apricots, nectarines, oranges, lemons, citrons, loquets (sic), guavas, cherries, cape, china and English mulberries, walnuts, Spanish chestnuts, almonds, medlars, quinces, grapes, pears, plums, figs, pomegranates, raspberries, strawberries and melons of all sorts, attain the highest degree of maturity in the open air...

W. C. Wentworth
The Colonial Kitchen Garden (1882)
(quoted in *An Australian Gardener's Anthology*)

FIG SURPRISES

tenderised dried figs
walnut or pecan halves
chopped crystallised ginger
icing sugar

Slit side of each fig and pop a walnut or pecan half in with a piece of crystallised ginger; press closed. Dust figs with sifted icing sugar and serve wlth coffee and liqueurs.

TURKISH FIGS

2-3 egg whites, well beaten
dried figs

rosewater
icing sugar, sifted

Place figs in bowl with rosewater to cover and soak overnight till figs have plumped up well. Drain. Dip figs in whisked egg white and then roll in icing sugar and place on greaseproof paper. Arrange on glass or silver platter with crystallised ginger and glacé fruits; serve with strong Turkish coffee.

A gift of wicker basket of fresh ripe figs, piled high in a pyramid on an arrangement of fig leaves, is particularly fortunate — it is said to signify a future legacy

Jams, Sauces and Preserves

Even though we now have the convenience of fridges and freezers, jewel-coloured jars of homemade jams and condiments are always welcome.

FIG AND GREEN GRAPE VINEGAR

1 kg/2 lb figs, sliced
1 kg/2 lb green grapes, stems removed
few tarragon leaves
cider vinegar, to cover

Place fruit and tarragon in wide-mouthed glass jars or preserving container and add sufficient cider vinegar to cover. Tie a muslin cloth over neck of jar and set aside in a dry, dark place for 2-3 weeks. Strain fruit and herbs and store in bottles, capping securely with non-metallic lids.

YUGOSLAVIAN FIG SAUCE

4 cups chopped peeled figs
½ cup water

juice of 1 lemon
2 tbspns capers, minced
2 tspns grainy mustard

Cook figs with water and lemon juice over medium heat until softened; pass mixture through a sieve or blend in food processor till smooth. Mix capers and mustard into fig purée and chill before serving with cold pork chicken or duck.

FIG AND CUMQUAT CHUTNEY

2 dozen firm, ripe cumquats
1 kg/2 lb pears, peeled, cored and chopped
250g/8 oz jellied ginger, finely chopped
2 tbspns powdered ginger
225ml/8 fl oz cider vinegar
175g/6 oz golden syrup
1 small onion, peeled and minced
6 dried figs, minced
2 tbspns raisins
1 tbspn powdered cinnamon
6-8 tbspns whisky
1 tspn ground cardamom
1 tbspn mustard seeds
½ tspn powdered cloves
water, to cover
brown sugar, to taste

Wash cumquats thoroughly, peel them, cutting the peel into long thin strips and reserving pith. Cut fruit into chunks, reserve seeds and place fruit and peel in large non-aluminium saucepan. Place pith and seeds in muslin bag, tie securely and put in saucepan with pears, jellied ginger, powdered ginger, cider vinegar, golden syrup, onion, figs, raisins, cinnamon, whisky, cardamom, mustard seeds and cloves. Add enough water to barely cover, place lid on saucepan and leave overnight.

Next day, bring mixture to the boil and simmer till peel is tender. Reduce heat and simmer, uncovered, for one hour or until mixture is pulpy. Measure mixture and add 1-1½ tbspns brown sugar for each cupful, or more or less to taste.

Bring to the boil to dissolve sugar and cook till jell point is reached. Spoon into heated sterile glass jars and cap securely.

FIG AND WATERMELON PRESERVE

This recipe is based on *Slatko od lubenice*, a traditional Yugoslavian delicacy.

2 cups white rind of watermelon, cubed
2kg/4 lb sugar
2 cups fig flesh, chopped
4 lemons, seeded and sliced
juice from 6 lemons
2 vanilla beans
pink watermelon flesh, seeded, puréed and sieved as necessary, to form 250ml/8 fl oz of juice

Place watermelon rind in large non-aluminium saucepan with cold water to cover and bring to the boil; cook until rind is quite tender, then drain and set aside. Make a syrup by boiling together sugar and 250ml/8 fl oz water. When syrup thickens, add watermelon rind, fig flesh, lemon slices, lemon juice, vanilla beans and watermelon juice. Continue to cook a further 15-20 minutes till mixture reaches desired consistency. Discard vanilla beans, spoon into warm sterile jars and cap securely.

TURKISH FIGS IN HEAVY SYRUP

Known as *Meyrah tatlush*, this sweet dish is offered to all guests by the woman of the house, both to signify the visitor is welcome and as a means of declaring her prowess as a cook—all Turkish women take pride in their ability to preserve fruit. Traditionally only a glass of water is offered with this dish, so the guest may fully savour the taste. Amongst Turkish men who are conducting business together, this sweet is usually served with strong black coffee.

1kg/2 lb small purple or green figs
1.25kg/2½ lb sugar
1½ cups water
juice of 2 lemons
1 tspn powdered cloves

Peel figs and cut off stem ends. With a needle, prick each fig twice. Combine sugar and water in a non-aluminium saucepan and bring to the boil, cooking rapidly to form a heavy syrup. Remove from heat and add figs, stirring once or twice and then setting aside for an hour. Return to heat and bring mixture to the boil until syrup thickens. Boil until syrup reaches stage where a drop placed in a glass of cold water will form a hard ball. Stir in lemon juice and ground cloves and cook for a further 1-2 minutes, stirring often, before spooning mixture into heated sterile jars.

INDIAN FIG CHUTNEY

1.75kg/3½ lb ripe tomatoes, peeled and chopped
2-3 cloves garlic, crushed
4 small green chillies, peeled, seeded and chopped
50g/1½ oz salt, or more to taste
4 ripe figs, peeled and sliced
150g/5 oz dried figs, chopped
200g/7 oz sultanas
300g/10 oz currants
4 ripe peaches, skinned, stoned and chopped
225g/7 oz ripe plums, skinned, stoned and chopped
450g/15 oz green cooking apples, peeled, cored and diced
550g/17 oz finely chopped almonds
225g/7 oz dried prunes, stoned and chopped
110g/3½ oz fresh ginger, peeled and grated
2 lemons, sliced thinly and seeded
1 tspn each cardamom and cayenne pepper
750ml/24 fl oz brown vinegar
325g/11 oz brown sugar

Place all ingredients in a large non-aluminium saucepan with vinegar to cover. Bring to the boil and then reduce heat and simmer for 1½-2 hours, or until mixture is desired pulpy consistency, adding more vinegar as necessary. Taste and adjust seasonings. Allow mixture to cool slightly before spooning into heated sterilised jars; cap securely. As with most chutneys and pickles, this recipe improves if left for at least 2-3 months before being opened. Refrigerate after use.

FIG AND CHESTNUT RELISH

1kg/2 lb fresh figs, peeled and puréed or mashed by hand
1 small onion, peeled and chopped
4 tbspns chestnut purée
1 tbspn chopped parsley
⅓ cup orange juice
zest from 1 lemon
1 tspn powdered cinnamon
½ tspn each powdered cloves and allspice
⅓ cup raspberry or cider vinegar
½ cup chopped hazelnuts
½ cup chopped walnuts

Combine all ingredients except nuts in a non-aluminium saucepan over low heat and gradually bring to the boil. Reduce heat and simmer for 30 minutes, adding extra vinegar if necessary. Add walnuts and hazelnuts and stir gently. Remove from heat and pour into heated sterile jars; cap securely.

Tip: for a rich stuffing for duck or turkey, combine 4-5 tbspns of this relish with 2 tbspns of cream.

SWEET FIG AND CORIANDER SAUCE

8 dried figs
2 tbspns sultanas, minced
1 tbspn fresh coriander, chopped
1 tbspn fresh mint, chopped
3 tbspns fresh lime juice
1 small red chilli, peeled, seeded and minced
2 cloves garlic, peeled and crushed
1 tspn chopped, peeled, fresh ginger
6 tbspns natural yoghurt
3 tbspns water
½ tspn clear honey
freshly ground black pepper and sea salt, to taste

Place figs in saucepan with enough water to cover; cover saucepan and simmer for 25 minutes, or until figs are tender. Drain, reserving liquid. Combine figs and all the other ingredients in food processor or mixing bowl, blending to form a paste. Add reserved water used to cook figs if a runnier consistency is required.

RUSSIAN FIG PRESERVES

In Russia, preserves like this are known as *Vareniye iz Inzhira*. They are offered to guests in winter time, along with plain cake or biscuits and a glass of unsweetened green tea.

1½kg/3 lb ripe purple figs, peeled and left whole
4 cups caster sugar
1 cup brown sugar
4 cups water
½ cup chopped walnuts
1/2 cup chopped hazelnuts
juice and zest of 1 lemon
juice and zest of 1 orange, peeled into long, julienne strips

Combine sugars and water in large non-aluminium saucepan, over moderate heat, stirring well until sugar dissolves. Add figs, nuts and orange and lemon juice and zest; reduce heat and simmer for 10 minutes. Remove from heat, cover and let stand for a few minutes before spooning into heated sterile glass jars. Cap jars securely and refrigerate.

SPICY FIG AND CINNAMON JAM

1.5kg/3 lb figs
500ml/18 fl oz water
100ml/4 fl oz cider vinegar
1 tspn cinnamon, or more to taste
2 tbspns dark rum
juice of 2 lemons
1 5cm/2″ cinnamon stick
½ tspn cloves
900g/2 lb white sugar

Skin figs and chop roughly. Combine in large, non-aluminium saucepan with water, vinegar, cinnamon, rum and lemon juice. Place cinnamon stick and cloves in small muslin spice bag, tie securely and add to pan. Bring to the boil and simmer for half an hour, mashing well against sides of pan. Remove muslin spice bag, bring to the boil again, and add sugar, stirring well until dissolved; boil rapidly for 1 hour or until jell point is reached. Spoon into heated, sterile jars and cap securely.

GRANDMOTHER'S FIG AND RHUBARB JAM

1kg/2 lb trimmed rhubarb, cut into 2.5cm/½″ lengths
500g/1 lb fresh figs, peeled
6 cups sugar
1 cup dried figs, minced
3 tbspns Grand Marnier
1 cup orange juice
grated zest and juice of 1 lemon
2 thin-skinned oranges
water, to cover

Place rhubarb, figs, sugar, dried figs, Grand Marnier, orange juice and grated zest and juice of lemon in bowl. Cover and set aside for 24 hours. Put oranges in a saucepan with enough water to cover and boil for 1 hour, or until skin becomes transparent. Let oranges cool slightly, then chop finely, discarding pips and combine with fig and rhubarb mixture in a heavy-based, non-aluminium saucepan over medium heat, stirring frequently until sugar has dissolved and most of the rhubarb has lost its shape. Bring to the boil and cook rapidly until setting point is reached. Remove from heat. Skim off any scum and allow to cool slightly before spooning into sterile, heated glass jars; cap securely.

FIG BUTTER

Use this delicious spread in place of marmalade or jam on hot crispy toast; or, lavish it on slices of rich fruit loaf.

4 cups fig pulp
1 cup sugar
¼ cup port wine
1 tbspn orange juice
1 tbspn grated orange zest
dash vanilla
2 tbspns unsalted butter

Place fig pulp in saucepan and cook over medium heat for 30 minutes, stirring constantly with a wooden spoon to stop it from sticking to the sides. Add sugar, wine, juice, zest and vanilla, and continue to cook rapidly for a further 15 minutes, stirring constantly. Remove from heat and stir through butter thoroughly. Spoon fig butter into heated sterile jars and cap securely.

Tip: Fruit 'leathers' or dried fruit strips are made using much the same method. To make fig 'leathers', omit the butter, spread the mixture out thinly on greaseproof paper and either dry in the sun or in a slow oven. Cut into slices and eat as a delicious snack.

SWEET FIG AND CHERRY PICKLE

300g/10 oz dried figs, chopped
350g/12 oz sugar
400ml/14 fl oz cider vinegar
4 tbspns sweet sherry
1 tspn each cloves, mace, ground ginger and mustard powder
1 tspn curry powder
250g/8 oz glacé cherries, halved

Soak figs overnight in enough cold water to cover; drain thoroughly. Combine sugar, vinegar, sherry and spices in a non-aluminium saucepan and bring to the boil. Reduce heat, add figs and simmer for 50 minutes; add cherries and simmer for a further 10 minutes before packing into sterile warmed jars. This is quite delicious with cold ham, pork or cold corned beef.

FIG AND PEPPERCORN PICKLE

500g/2 lb fresh figs, peeled and chopped
½ cup sugar
1¼ cups cider vinegar
¾ cup orange juice
250g/8 oz jellied ginger, finely chopped
1 tbspn whole cloves
good pinch salt
1 tbspn powdered cinnamon

2 tspns whole allspice berries
I tspn each green and pink peppercorns
I tbspn brandy

Pick over and prepare figs. Combine with other ingredients in a large, non-aluminium saucepan. Simmer, covered, over low heat for 25 minutes. Remove from heat and allow to cool. Spoon into warmed, sterile jars and cap securely. This pickle is delicious served with cold ham or pork.

WARM FIG AND BERRY SAUCE

2 quinces, skinned and diced
3 tbspns caster sugar
juice and zest of I lemon
2 punnets of strawberries, washed, hulled and halved
I persimmon, skinned and diced
I tbspn each orange juice and apple juice
2 medium figs, halved
I tbspn port wine

Place quince pieces in bowl with sugar, lemon juice and zest; mix well, cover and refrigerate overnight. Place in a saucepan and simmer gently till fruit is soft. Allow quince mixture to cool slightly, then place in food processor with strawberries. Lightly sauté persimmon pieces in orange and apple juice and reduce until liquid nearly disappears. Allow to cool slightly and add to mixture in food processor with port wine; blend till smooth. Scoop out fig flesh and break up with a fork; fold through sauce and serve warm with vanilla ice cream.
Tip: For an extra special effect, fold three tbspns of whipped cream through sauce.

CRUNCHY CARROT AND FIG RELISH

500g/I lb carrots, peeled, cooked and puréed
2 large carrots, peeled and coarsely grated
3 tbspns minced, dried figs
2 tbspns raisins
2 cups chopped brown onion
¾ cup raw sugar
juice of 2 oranges

Figs, a heavenly delight.

. . . figs may have been cultivated in an espaliered form in the elegant walled gardens which formed part of the Pharaoh's court . . .

zest of 1 lemon
3 tspns black peppercorns
2 tspns cumin
2 tspns turmeric
⅓ cup walnuts, roughly chopped
2 tbspns brandy

In a large non-aluminium saucepan, combine all ingredients except brandy. Bring to the boil, reduce heat and simmer for 15-20 minutes. Stir in brandy and cook for a further 2-3 minutes. Spoon into warmed, sterile glass jars and cap securely.

JELLIED FIG AND WALNUT RELISH

A delicious accompaniment to cold meats, especially cold roast lamb.

2 tbspns gelatine
½ cup cold water
¼ cup port wine
¼ cup reserved mandarin syrup
sugar, to taste
1 medium orange, peeled, seeded and chunked
500g/1 lb mandarin segments, drained (reserve syrup) and chopped
½ cup walnuts, coarsely chopped
½ cup dried figs, minced

Combine gelatine and water in blender jug and let stand. Heat port wine and mandarin syrup and add to mixture, processing at low speed until gelatine dissolves. Add sugar and orange and blend to a purée. Add mandarin segments, walnuts and figs and stir through gently. Decant in glass serving dish(es) and chill until firm.

FIG AND PINEAPPLE JAM

2 kg/4 lb fresh figs, peeled and chopped
1 kg/2 lb chopped, cored, fresh pineapple
225g/7½ oz preserved ginger in syrup, chopped
550ml/19 fl oz water

3kg/6 lb white sugar
6 tblspns lemon juice

Combine figs, pineapple, ginger and water in a large non-aluminium saucepan over moderate heat and simmer until fruit is pulpy. Add sugar, stirring well till dissolved, then add lemon juice. Bring jam to the boil and cook uncovered until jell point is reached. Allow to cool slightly, then spoon into sterilised, heated jars and cap securely.

SWEET FIG AND CHILLI CHUTNEY

2 cups dried figs
1½ cups sultanas
¾ cup brown sugar, firmly packed
1 cup white wine vinegar
125g/4 oz preserved ginger, chopped
3 small dried red chillies, seeded and finely chopped
juice of 1 orange
1½ tspns mustard seeds
1 tspn salt, or to taste
1 cup water, as required

Place dried figs and sultanas in china bowl with enough boiling water to cover and set aside; leave overnight, then drain. In large, non-aluminium saucepan, combine sugar, vinegar, preserved ginger, chillies, orange juice, mustard seeds, salt and water. Bring to the boil and simmer, uncovered, until mixture thickens (about 20 minutes). Stir in figs and sultanas and continue to simmer a further 20 minutes until fruit softens. Spoon into heated sterile jars and cap securely.

FIG AND HONEYDEW MELON JAM

2kg/4 lb honeydew melon, peeled, seeded and chopped
2kg/4 lb figs, peeled and sliced
3kg/6 lb sugar
juice and zest of 5 oranges
juice and zest of 10 lemons

Place melon and figs in large, non-aluminium saucepan and sprinkle with 2-3 tbspns of the sugar; cover and leave overnight.

Bring mixture to the boil uncovered, then reduce heat and simmer till fruit is soft and pulpy. Add lemon and orange juices and zests and stir through mixture; add sugar slowly, continuing to mix well until sugar is dissolved. Continue to cook over low to moderate heat until desired consistency is reached. Allow to cool slightly before pouring into heated, sterilised jars; cap securely.

FIG FONDUE SAUCE

2 punnets strawberries, washed, with stems retained
150ml/5 fl oz sweetened condensed milk
250g/8 oz dark cooking chocolate, grated
⅓ cup heavy cream
6 dried figs, minced
2 tspns instant coffee or Jamaican coffee essence
¼ cup cognac

In a double saucepan over low heat, combine condensed milk, chocolate and cream. Stir in figs, coffee and brandy, keeping heat very low and constant. Transfer to bowl and serve warm, using whole strawberries to dip into mixture.

Tip: For an extra-delectable taste sensation, substitute 250g/8 oz Toblerone milk chocolate for the cooking chocolate.

FIG AND MINTED YOGHURT DIP

2 tbspns fresh mint, minced
1 tbspn fresh coriander, minced
200g/7 oz natural yoghurt
1 tbspn minced, dried figs
½ tspn minced, fresh ginger
½ tspn each chilli powder and garam masala
dash lime juice
½ tspn clear honey (optional)

Combine all ingredients in a bowl, whisking till smooth; chill. Serve with flat *pita* bread and crudités.

FIG AND APPLE SPREAD

This is delicious on pancakes and muffins, or in sandwiches;

also, try it as a porridge topping when the weather is a little cooler.

125g/4½ oz dried figs, finely chopped
125g/4½ oz dried apples, finely chopped
1 cup unsweetened orange juice
1 cup unsweetened pineapple juice
½ lemon, pipped and thinly sliced

Combine all ingredients in a large non-aluminium saucepan. Cover and cook until liquid is reduced and fruit is well softened and pulpy. Purée for a smooth consistency or leave as is for a chunkier texture.

FIG AND PEACH SPREAD

100g/3½ oz dried figs, chopped
150g/5 oz dried peaches, chopped
100g/3½ oz dates, stoned and chopped
100g/3½ oz dried apricots, chopped
grated zest of 1 orange
pinch each nutmeg and cinnamon
250ml/8 oz orange juice
100ml/3½ fl oz grape juice

Combine all the ingredients in a non-aluminium saucepan over moderate heat. Bring to the boil, then reduce heat and simmer until liquid is reduced and mixture starts to thicken. Stir thoroughly with a wooden spoon until mixture is smooth and then pour into warmed, sterile glass jars; cap securely.

FIGS IN COGNAC

1.5kg/3 lb purple figs, washed and quartered
zest and juice of 2 oranges
4 cups sugar
1 cup water
4 cups cognac

Combine figs in a non-aluminium saucepan with orange zest and juice and cook gently until fruit has softened slightly;

On patience

The philosopher Epictetus, who lived from 50 to 120 AD, wrote:

'Nothing great is created suddenly, any more than a bunch of grapes or a fig. If you tell me that you desire a fig, I answer you that there must be time. Let it first blossom, then bear fruit, then ripen.'

remove from heat. In a separate pan, boil sugar and water to form a syrup. Put figs into a bowl and pour syrup over. Let stand overnight.

Transfer fig and syrup mixture to a saucepan. Bring to the boil and cook gently for 8-10 minutes; drain and reserve syrup and pack figs into sterile heated jars. Reheat syrup and bring to the boil, stirring rapidly until it thickens. Stir through cognac and pour over figs; cover and cap securely.

FIG AND PLUM SAUCE

2kg/4 lb ripe plums, stoned, skinned and chopped
1kg/2 lb ripe figs, peeled and chopped
2 large Spanish onions, peeled and minced
2 tspns soy sauce
3 cups malt vinegar
1 small hot green chilli, seeded and minced
2 tspns mustard seeds
½ stick cinnamon
¾ cup brown sugar
1 tspn ground mace
½ tspn ground cloves

Combine plums, figs and onion in food processor, blending on low speed till quite smooth. Put plum purée in large, non-aluminium saucepan with soy sauce and half the vinegar. Tie chilli, mustard seeds and cinnamon stick securely in a piece of cheesecloth or muslin and add to purée mixture. Add sugar, mace and cloves and cook over medium heat for 10-15 minutes, stirring often until mixture becomes pulpy. Add rest of vinegar and continue to cook until mixture reaches desired consistency. Discard spice bag and allow to cool slightly, before spooning into heated sterile glass jars; cap securely.

Figs for Something Completely Different

It was the famous Roman cook and philosopher Apicius who, in the first century AD, devised a spicy stuffing for poultry made from onions, walnuts, dried figs and apricots. In Middle Eastern and European countries, meat and vegetable dishes are still often enhanced with dried or fresh fruits, especially figs, nuts and sweet spices. Many traditional meat dishes lend themselves to unusual but compatible sauces or stuffings based on figs.

CHICKEN CASABLANCA WITH FIG SYRUP

1.5kg/3 lb roasting chicken
6 dried figs, quartered
handful dried apricots, chopped
1 green apple, peeled, cored and chopped
½ tspn each cinnamon and ginger
pepper
oil or butter, for roasting

Fig Syrup

6-8 dried figs, chopped
3-4 dried apricots, chopped
1 cup chicken stock
½ cup orange juice
¼ cup sherry
1 tbspn honey
½ tspn ground ginger

Set oven to 190°C/375°F. To prepare chicken, combine figs, apricots, apple, cinnamon, ginger and pepper in bowl and stuff cavity with mixture, securing with poultry pins or skewers. Place chicken in baking dish with 2 tbspns oil or a knob of butter, to preference, and roast for 1½-2 hours, basting often with pan juices.

For syrup, combine figs and apricots with other ingredients over low heat for 30 minutes, or until fruit is well softened. Process mixture in blender till smooth and return to heat, simmering until chicken is ready. To serve, drizzle syrup over slices of chicken and stuffing.

FRUITY TOFU STIR-FRY

12 shallots, blanched till tender
6 tiny, yellow squash, blanched and halved
2 small Japanese eggplants, blanched and sliced lengthwise
½ red or yellow capsicum, seeded and chopped
1 orange, peeled, seeded and chopped
½ cup dried figs, finely chopped
2-3 tbspns walnuts, chopped
350g/12 oz tofu, cubed
2 tbspns soy sauce

I tbspn coarse grain mustard
2 tbspns rice wine
I tspn finely minced fresh ginger
I tbspn honey
3 tbspns orange juice
I tbspn each minced fresh parsley and mint
freshly ground pepper, to taste
2 tbspns peanut oil

Combine shallots, squash, eggplant, capsicum, orange, figs, walnuts and tofu in bowl. In separate bowl, blend soy sauce, mustard, rice wine, ginger, honey and orange juice. Stir in herbs and pepper, to taste. Spoon marinade over vegetables and fruit and let stand for 15 minutes or longer. Heat oil in large frying pan or wok and lightly sauté mixture for 3-5 minutes, or until tofu is warmed through. Serve with brown rice.

FIG AND FLOWER SALAD

⅓ cup currants
4 fresh purple figs, quartered
⅓ cup pine nuts
2 punnets cherry tomatoes, halved
6 shallots, finely sliced
½ cup walnut oil
3 tbspns cider vinegar
I tbspn orange juice
I tbspn muscat liqueur
salt and freshly ground pepper
⅓ cup violets, washed and patted dry

Combine currants, figs, pine nuts, tomatoes and shallots in a bowl. In a separate lidded container, blend walnut oil, vinegar, orange juice, muscat and salt and pepper and shake well; pour over fig mixture and serve, garnished with violets.

FIG AND RASPBERRY SALAD

½ cup peeled, sliced, fresh peaches
2 cups fresh raspberries
I cup sliced fresh figs

1 cup peeled, sliced kiwi fruit
1 cup peeled, chopped mango
½ cup fresh blackberries
¼ cup lime juice (fresh preferably)
½ cup grape juice
dash each orange juice and Curaçao
few fresh mint leaves, to garnish

Combine all ingredients in a large china bowl, tossing well; serve, garnished with mint, in individual wine glasses.

MOROCCAN STUFFED APPLES

1 tbspn butter
1 chicken breast, skinned and diced
2 tbspns chopped pine nuts
1 tspn ground ginger
¼ tspn each powdered cinnamon and cloves
⅓ cup chopped dried figs
⅓ cup chopped dried peaches
⅓ cup currants
6 medium sized green cooking apples, cored

Melt butter in a frypan and cook chicken and pine nuts, ginger, cinnamon and cloves for 5 minutes. Remove from heat and stir through figs, peaches and currants. Core apples and stuff cavities generously with fruit and chicken mixture. Place in a baking dish and add water to the depth of one inch up the sides of the apples. Bake in a moderate oven for 45 minutes, or until quite tender.

INDIAN FIG SALAD WITH LIME DRESSING

8 fresh figs, quartered lengthwise
4 pink grapefruit, peeled with pith removed and broken into segments
1 ripe nectarine or peach, stoned and diced
100g/3½ oz seedless green grapes, halved
juice of 2 fresh limes
4 tbspns clear honey
pinch each cardamom and cinnamon
1 tspn minced fresh ginger root
natural yoghurt, to garnish

Gently mix together figs, grapefruit, nectarine or peach and grapes. Combine lime juice, honey, cardamom, cinnamon and ginger in lidded jar and shake well to mix. Pour over fruit and stir gently again; serve immediately.

FIG FRUIT MEDLEY

4 fresh purple or green figs, washed and quartered
1 cup passionfruit pulp
1½ cups freshly squeezed orange juice
½ cup blueberries
1 mango, peeled, stoned and cut into chunks
¼ cup peeled, chopped pawpaw
3 kiwi fruit, peeled and sliced

Combine figs, passionfruit and orange juice in saucepan over low heat for a few minutes. Remove from heat and pour over rest of fruit. Mix lightly and serve immediately, slightly warm, over mounds of ice cream or fresh fruit sorbet.

FIGS AND PROSCIUTTO HAM

A simple, time-honoured and delicious favourite, the perfect marriage of saltiness and sweetness.

12 ripe purple figs
250g/8 oz finely sliced prosciutto ham (parma ham)
freshly ground black pepper
fresh fig leaves, if available

Wash figs, cut in quarters back to the stem and fold out to form a 'flower'. Arrange figs with thin slices of prosciutto on fresh fig leaves in a shallow dish or platter. Chill until ready to serve, then serve immediately, offering pepper grinder to accompany.

FIG, BANANA AND TAHINI SMOOTHIE

½ cup carbonated fruit juice or combination fruit juice and carbonated mineral water—pineapple or apple would be good
½ cup soy milk
1 large banana, peeled and chopped

2 tbspns tahini
4 dried figs, chopped
pinch nutmeg
fresh fruit in season (eg: kiwi fruit, peaches, cherries), to garnish

Process all ingredients in blender until smooth. Garnish with fresh fruit and serve immediately.

FRESH FIGS WITH GINGERED RICOTTA CHEESE AND RASPBERRY SORBET

8 fresh figs
150g/5 oz ricotta cheese
50g/1½ oz crystallised ginger, finely diced
300g/10 oz raspberry sorbet
3-4 tbspns kiwi fruit pulp
few sprigs mint

Halve figs and scoop out flesh, retaining skins. Combine fig flesh gently with ricotta cheese and ginger, using a fork; set aside. Place a scoopful of raspberry sorbet in the centre of serving dish(es) and arrange fig shells decoratively around sorbet. Fill each fig shell with cheese mixture, mounding it up generously; garnish each with a drizzle of kiwi fruit pulp and decorate with mint sprigs.

SPICY FIG, APPLE AND HONEY SOUP

A delicious summertime soup that makes the most of sweet fragrant spices.

1.5kg/3 lb green apples, peeled, cored and chopped
170g/6 oz dried figs, chopped
1 tbspn raisins
225ml/8 fl oz apple juice
625ml/1 pint water
2 tbspns marsala
1 tbspn brandy
3-4 tbspns clear honey, to taste
3 tbspns fresh lemon juice
1 tspn each ground cinnamon and allspice

pinch each cloves and ground ginger
½ tspn each freshly grated nutmeg and ground cardamom
1 tspn vanilla, to taste
1 pint buttermilk
200ml/8 fl oz natural yoghurt
slivered almonds, thinly sliced lemons and extra cinnamon, to garnish

Combine apples, figs and raisins with apple juice, water, marsala, brandy, honey, lemon juice, spices and vanilla in a large non-aluminium saucepan over gentle heat. Cover and simmer for 35-40 minutes, until fruit is softened. Purée mixture in blender or food processor; transfer to serving bowl and allow to cool slightly before stirring buttermilk and natural yoghurt through. Cover and chill well in refrigerator. Taste and adjust seasonings, adding extra cinnamon or honey, if desired. Serve soup in chilled bowls, garnished with almonds, lemon slices and extra cinnamon.

SAVOURY CREAMED FIGS

12 ripe figs, quartered lengthwise

Dressing

150ml/5 fl oz natural yoghurt
150g/5 oz ricotta cheese
1 tbspn tahini spread
1 tbspn clear honey
2 tbspns lemon juice
1 tbspn light vegetable oil
freshly ground black pepper
fresh coriander leaves, to garnish

Arrange figs in serving dish(es) and chill. Place ingredients for dressing in food processor and blend till smooth; chill. Pour dressing over figs, garnish with coriander and serve immediately.

ROAST CHICKEN WITH FIG AND MUSCAT SAUCE

This recipe dates from Roman times when people loved to eat meat mixed with fruit.

2kg/4 lb roasting chicken with neck and giblets
juice of large orange
225g/7 oz dried figs, chopped finely
225g/7 oz hazelnuts, shelled and crushed
1 large onion, peeled and chopped
1 tbspn grated lemon zest
60g/2 oz breadcrumbs
250g/8 oz butter, melted
salt and freshly ground pepper

Sauce
100g/3½ oz butter
1 stalk celery, chopped
1 tbspn flour
½ tspn mustard powder
1 cup muscat liqueur
1 tbspn lemon juice
1 tspn honey
½ cup dried muscatel grapes
½ cup chopped dried figs

Preheat oven to 190°C/375°F. Wash and dry chicken. In a bowl, mix together orange juice, figs, hazelnuts, onion, lemon zest and breadcrumbs and spoon mixture inside bird, securing cavity with poultry pins. Place chicken in baking dish, drizzle melted butter over and season with salt and pepper. Roast for 1¼-1½ hours, basting frequently, or until juices come away clear.

To make sauce, melt butter and cook celery over medium heat. Stir flour and mustard powder into butter and mix till pasty, then add muscat, lemon juice and honey; cook gently for 20-25 minutes. Add grapes and figs and cook for a further 20 minutes, adding extra muscat if desired and salt and pepper, to taste.

GINGERED DUCK WITH FIGS

1.5kg/3 lb duck
30g/1 oz butter
5 shallots, finely chopped
1 tbspn orange zest

2 tspn tomato paste
¾ cup chopped dried figs
½ cup dried apricots
1 cup dried figs
¾ cup muscatels
1 cup apple cider
½ cup green ginger wine
1 ½″ piece root ginger, peeled and grated
1 tspn cinnamon
½ tspn powdered cloves
2 tbspns flour
¼ cup sherry

Set oven at 180°C/350°F. Wash duck and pat dry inside and out. Combine butter, shallots, orange zest, tomato paste and ¾ cup dried figs and place in cavity of duck; secure with poultry pins. Place on a wire rack in a pan and roast in moderate oven for 1 hour or until cooked.
Meanwhile, combine apricots, figs, muscatels, apple cider, green ginger wine, ginger, cinnamon and cloves in pan, bring to the boil and then simmer till fruit is soft. Strain fruits and reserve liquid. When duck is cooked, place on serving platter and decorate with fruits. Drain away all but 1 tbspn fat from baking dish and add flour, stirring constantly until it is well browned. Add reserved fruit liquid and bring to a simmer. Stir in sherry. Spoon fig and ginger sauce over duck and serve remaining sauce separately. If desired, garnish with two or three halved figs.

PORK WITH FIGS AND APPLES

This is based on a medieval recipe—in those times, fruits were often used to enhance the flavour of meat. The acid in the apples and figs also helps to make the pork tender and more digestible.

2 leeks, peeled and sliced
900g pork loin chops, trimmed
1 cup light stock
1 cup white wine (preferably sweetish)
1 tbspn white wine vinegar

1 tbspn white wine vinegar
½ tspn each powdered cinnamon and cumin
2 tbspns chopped parsley
salt and freshly ground black pepper, to taste
450g/15½ oz cooking apples, peeled, cored and sliced
125g/4 oz dried figs, chopped

Fry leeks and pork in a little butter in a large heavy-based cooking pan for 5 minutes, or until leeks are transparent and meat is well sealed. Add stock, wine, vinegar, honey, coriander, cumin, parsley and salt and pepper to taste. Simmer gently for 5 minutes, then add apples and figs. Continue cooking until meat is tender, adding a little extra honey and wine, if required. Serve with boiled rice and lightly steamed zucchini strips.

FIG AND HIBISCUS TEA COMPOTE

A delicious and refreshing breakfast dish, or any-time snack.

150g/5 oz dried figs, chopped
100g/3½ oz sultanas
100g/3½ oz dried apricots, chopped
100g/3½ oz prunes, stoned and chopped
grated zest and juice of 1 lemon
grated zest and juice from 2 oranges
juice of 1 lime
250ml/8 fl oz cooled hibiscus tea, strained
sour cream or natural yoghurt, to serve

Combine figs, sultanas, apricots and prunes in large china bowl. Mix in lemon and orange zest and juice and lime juice. Pour over this the cooled, strained hibiscus tea, cover and refrigerate overnight, stirring occasionally. Serve chilled with a dollop of sour cream or yoghurt.

MEDIEVAL MEAT BALLS

These were once called 'golden apples', due to the colouring effect of the saffron.

good pinch saffron
2 tbspns warm milk
125g/14 oz soft breadcrumbs
500g/1 lb lean pork mince
1 onion, peeled and minced
1 tspn chopped mint
2 tbspns minced dried figs
1 tbspn minced dried apricots
1 tbspn sultanas
freshly grated nutmeg
salt and freshly ground pepper, to taste
2 eggs well beaten
flour
oil
lemon wedge, to garnish

Allow saffron to dissolve in 1 tbspn of the milk and set aside. Mix remaining milk into breadcrumbs, then combine with pork, onion, mint, figs, apricots, sultanas, nutmeg, salt and pepper and eggs; mix well. Stir in saffron liquid. Shape mixture into balls and dip into flour before frying in hot oil till golden. Serve with lemon wedges.

Tip: For a truly authentic touch, these meat balls should be threaded onto a long spit—or sword!—and cooked over an open fire.

SWEET VEAL PIE

Pies such as this, well spiced and sweetened with fruit, were perhaps the forerunner of today's shepherd's pies. Certainly, in many early household almanacs, 'receipts' for meat pies which also contained fruit such as figs, pears and quinces, were common. This recipe is adapted from a 16th century recipe.

650g/1½ lb veal trimmed and cubed
3 tbspns flour
300g/10 oz shortcrust pastry
110g/3½ oz dried figs, chopped
4-6 plums, stoned and diced

1 tbspn raisins
1 tbspn chopped candied peel
salt and freshly ground black pepper
½ tspn each powdered mace and grated nutmeg
pinch each cinnamon and powdered cloves
2 tbspns sweet dark rum
melted butter, for brushing over pie

Roll veal in flour and fry in a little butter to seal in juices; drain and set aside. Preheat oven to 180°C/350°F. Roll out half the pastry to form a circular lining for pie dish and press into dish, trimming edges and reserving trimmings. Brush with melted butter, then fill pie shell with veal, figs, plums, raisins and peel. Mix spices with rum and pour over fruit and veal mixture. Roll out remaining pastry and cut to form a lid for pie, pinching edges together and trimming with a knife. Brush top of pie with remaining butter. Bake in the oven until pastry is golden (30-40 minutes).

ELIZABETHAN ROAST DUCK WITH FRUIT STUFFING

2 kg duck
1 tbspn flour
1 tspn salt
1 tspn powdered ginger

Stuffing

1 pear, peeled, cored and chopped
1 apple, peeled, cored and chopped
2 plums, peeled, stoned and chopped
2 figs, peeled and chopped
small bunch seedless sultana grapes
½ cup toasted slivered almonds
1 tspn sage
1 tspn cumin
salt and pepper, to taste

Preheat oven to 220°C/450°F. Wash and dry duck and pierce skin all over with a fine skewer. Mix together flour, salt and ginger and rub over duck. To make stuffing, combine pear, apple, plums, figs, grapes, almonds, sage, cumin and salt and pepper to taste in bowl and mix well. Stuff into cavity of bird and secure with poultry pins. Place in roasting pan and cook for 20 minutes before reducing heat to 180°C/350°F; continue to cook, allowing 30 minutes per 450g/1 lb weight, pouring off fat and basting with reserved pan juices. Serve with wild rice, baby carrots and lightly steamed asparagus.

Being an ancient fertility symbol, it is unsurprising that figs were often baked in cakes or pies given to Mother on 'Mothering Sunday', the English precursor to the more commercialised Mother's Day we still celebrate today. This 'Mothering Sunday' fig pie is based on a traditional 18th century recipe; it would probably have been decorated with sugar eggs, flowers and leaves and have had the words 'Mother' or 'Mothering Sunday' piped in icing, and the edges of the pastry shell would have been scalloped or twisted in an especially decorative way.

a pre-cooked pastry shell
2 cups dried figs, soaked overnight
¼ cup currants
¾ cup sugar
1 tbspn treacle
½ tspn mixed spice
1 tbspn orange rind
2 egg whites

Chop figs into small pieces. Add currants, sugar and treacle, spices and rind. Beat egg whites till stiff; fold these into the fig mixture and mix well. Pour mixture into pastry shell and bake in a medium oven 190°C/375°F) for about 20 minutes or until golden brown. Serve cold with whipped cream.

Bibliography

The Greek Cookbook, Tess Mallos, Summit Books, Paul Hamlyn Sydney 1976

The Complete Cookery Course, C. Ellwood and R. Wadey, Octopus Books, London 1986

Origins of Festivals and Feasts, Jean Harrowven, Kaye and Ward, London 1980

The Encyclopaedia of Superstitions, H. Robinson Spencer, Kaye and Ward, London 1962

Haute Cuisine, Heart Cuisine, The International Olive Oil Council/Heartwatch joint promotional, publication, Australia 1991

An Australian Gardener's Anthology, McPhee Gribble Publishers, Rigby Publishers, Adelaide 1982

Traditional Herb and Spice Cookery, Jack Santa Maria, Rider and Company/Century Hutchinson, London 1985

Traditional Australian Cooking, Shirley Constantine, McPhee Gribble/Penguin, Australia 1991

The Balkan Cookbook, Inge Kramarz, Crown Publishers, New York 1972

Tastes of the Sun, E. Chambers, Murdoch Magazines, Sydney NSW (no date)

Kitchen in the Kasbah, Irene F. Day, Andre Deutsch Ltd., London 1975

Fruit Cakes, Lois Stocks, McCulloch Publishing, Carlton Vic., 1989

Carol, Gaily Carol, ed. B. Harrop, A&C Black, 35 Bedford Row, London, WC1R4JH

Grandma's Preserving Secrets, P. Hutchinson, W. Foulsham Co., London 1978

The New Vegetarian Epicure, Anna Thomas, Penguin, London 1978

The New World Encyclopaedia of Cooking, The Culinary Arts Institute, North American Education Guild, Manitoba, USA 1973

Preserving with Fowlers Vacola, Ann Creber, Regency Publications, Victoria, Australia 1988

Great Desserts and Pastries, Robert Carrier, Angus & Robertson, Australia 1979

The Wholefood Book, G. Seddon and J. Burrow, Mitchell Beazley Publishers, London 1978

Homemade — at a fraction the cost, Polly Pinder, Search Press, Kent UK 1983

A Taste for all Seasons, B. Sutherland Smith, Lansdowne Press, Sydney, Australia 1975

New Ways to Eat Well, S. Tomnay, Bay Books, Sydney NSW (no date)

Margaret Fulton's Book of Soups and Starters, Octopus Books, London 1983

Pure, Fresh and Simple, R. Jensen, Greenhouse Publications, Victoria 1987

The Fresh Fruit Cookbook, V. Bacon, Regency Publications, Prahran Vic. (no date)

Fruit for the home and garden, L. Johns and V. Stevenson, Angus & Robertson, Sydney NSW 1979

Growing Tree Fruits, B. Bazeley, William Collins and Sons Ltd., London 1990

The Time is Ripe, G. Wilson, Viking O'Neil/Penguin Books, Australia 1988

Feasting on Raw Foods, ed. C. Gerras, Rodal Press, USA 1980

Margaret Fulton's Book of Preserves and Pickles, Octopus Books, London 1981

Food with the Famous, Jane Grigson, Hollen Street Press, London 1979
Beverley Sutherland Smith's Collector's Edition, The Five Mile Press, Victoria 1987
The Great Cooks of Australia Cookbook, Macmillan, South Melbourne, 1985
Martha Stewart's Quick Cook Book, Sidgwick and Jackson, London 1983
Robert Carrier's Quick Cook Book, The Hamlyn Publisher Group, London 1983
Fruit and Vegetables of the World, John Goode, Ross Publishing/Lothian Publishing, Melbourne 1987
The Good Old-fashioned Cookbook, Clare Stephens, Lansdowne Press/Wellington Newspapers, Australia 1981
Harvest Cookbook, Mark Pearson, Harvest Publishing Balmain, Sydney 1981
A Gourmet's Book of Fruit, J. Bastyra and J. Canning, Golden Press, Drummoyne Australia 1989
Encyclopaedia of Fruit and Cookery, Margaret Fulton, Octopus Books, Australia 1983
Harrods Cookery Book, M. Aslani, Angus & Robertson, Sydney 1985
The Fruit Cookbook, S. Jonas, Reed Books, Frenchs Forest NSW 1985
The British Museum Cookbook, M Berriedale-Johnson, British Museum Publications Ltd., London 1987
California Cooking, Mabel and Gar Hoffman, HP Books, USA 1983 (Fisher Publishing)
Seasonal Gifts from the Kitchen, E. Crumpacker, Wm Morrow and Co., Inc., New York 1983
The Australian Womens Weekly Gardening Book, Alan Seale, Golden Press, Potts Point Australia 1971
Elise Pascoe's Four Seasons of Food and Wine, Macmillan Company, South Melbourne 1984
The Italian Store Cupboard, Elizabeth David, Macdonald, Great Britain 1954
The In Season Cookbook, Nancy Baldwin, Hyland House, South Yarra Melbourne 1989
The Kitchen Handbook, Readers Digest Services Pty Ltd, NSW 1982
Jane Grigson's Fruit Book, Penguin, Great Britain 1982
Food, Wvaelery Root, Simon & Schuster, New York 1980
What Food is That¿, Jo Rogers, Weldon Publishing, Australia 1990
The Terrace Times Cookbook, Helen Arbib, The Terrace Times, Paddington, NSW 1987
The Whole World Cookbook, J. Killeen, Charles Scribner's Sons, USA 1979
The Good Housekeeping Cookbook, The Hearst Corporation, USA 1973
The Fruit and Nut Book, H. Radecka, McGraw-Hill/Phoebe Phillips/Spain 1984
Fruit for the Home and Garden, L. Johns and V. Stevenson, Angus & Robertson, Sydney 1985
The Alcoholic Cookbook, J. Stone, Michael Joseph, Great Britain 1972
A Country Compendium, Godfrey Baseley, Sidgwick and Jackson, Great Britain 1977

The Hearst Corporation, USA 1973
The Fruit and Nut Book, H. Radecka, McGraw-Hill/Phoebe Phillips/Spain 1984
Fruit for the Home and Garden, L. Johns and V. Stevenson, Angus & Robertson, Sydney 1985
The Alcoholic Cookbook, J. Stone, Michael Joseph, Great Britain 1972
A Country Compendium, Godfrey Baseley, Sidgwick and Jackson, Great Britain 1977

Index to Recipes